# 陶瓷迹型学概论

周勇　周强／著

SURVEY OF
CERAMICS
TRACE MODEL
STUDY

中国·广州

## 图书在版编目（CIP）数据

陶瓷迹型学概论 / 周勇, 周强著. — 广州 ：南方日报出版社，2013.5
ISBN 978-7-5491-0829-9

Ⅰ．①陶… Ⅱ．①周… ②周… Ⅲ．①陶瓷－工艺学－研究 Ⅳ．①TQ174.1

中国版本图书馆CIP数据核字(2013)第085726号

TAOCI JIXINGXUE GAILUN
**陶瓷迹型学概论**　　　　　　　　　　　　　　　　　　　周勇　周强 著

| | |
|---|---|
| 出版发行： | 南方日报出版社 |
| 地　　址： | 广州市广州大道中289号 |
| 电　　话： | (020) 83000502 |
| 经　　销： | 全国新华书店 |
| 印　　刷： | 广东信源彩色印务有限公司 |
| 开　　本： | 889mm×1194mm　1/16 |
| 印　　张： | 17.5 |
| 字　　数： | 650 千字 |
| 版　　次： | 2013年5月第1版 |
| 印　　次： | 2013年5月第1次印刷 |
| 定　　价： | 200.00元 |

投稿热线：(020) 83000503　　读者热线：(020) 83000502

网址：http://www.nfdailypress.com/

发现印装质量问题，影响阅读，请与承印厂联系调换

# 陶瓷迹型研究系列丛书

《陶瓷迹型学概论》
《元青花迹型研究与鉴定》

　　　　　　　　　　　　　　　　　　　　计划2013年出版

《康熙青花及五彩迹型研究与鉴定》
《汉代陶瓷迹型研究与鉴定》

　　　　　　　　　　　　　　　　　　　　计划2014年出版

《唐三彩迹型研究与鉴定》
《永宣青花迹型研究与鉴定》
《越窑青瓷迹型研究与鉴定》

　　　　　　　　　　　　　　　　　　　　计划2015年出版

《耀州窑瓷器迹型研究与鉴定》
《成化青花及斗彩迹型研究与鉴定》

　　　　　　　　　　　　　　　　　　　　计划2016年出版

《龙泉窑瓷器迹型研究与鉴定》
《宋代湖田窑瓷器迹型研究与鉴定》

　　　　　　　　　　　　　　　　　　　　计划2017年出版

《吉州窑瓷器迹型研究与鉴定》
《建窑瓷器迹型研究与鉴定》

　　　　　　　　　　　　　　　　　　　　计划2018年出版

# The Ceramics Trace Model Study Series

Survey of Ceramics Trace Model Study
Trace Model Study and Authentication: Yuan Dynasty Underglaze Blue Porcelain
Planned Date of Publishing: 2013

Trace Model Study and Authentication: Kangxi Underglaze Blue and Wucai Porcelain
Trace Model Study and Authentication: Han Dynasty Ceramics
Planned Date of Publishing: 2014

Trace Model Study and Authentication: Tang Dynasty Three Color Ceramics
Trace Model Study and Authentication: Yongle and Xuande Underglaze Blue Porcelain
Trace Model Study and Authentication: Yue Kiln Celadon Wares
Planned Date of Publishing: 2015

Trace Model Study and Authentication: Yaozhou Kiln Porcelain
Trace Model Study and Authentication: Chenghua Underglaze Blue and Doucai Porcelain
Planned Date of Publishing: 2016

Trace Model Study and Authentication: Longquan Kiln Porcelain
Trace Model Study and Authentication: Song Dynasty Hutian Kiln Porcelain
Planned Date of Publishing: 2017

Trace Model Study and Authentication: Jizhou Kiln Porcelain
Trace Model Study and Authentication: Jian Kiln Porcelain
Planned Date of Publishing: 2018

# 陶瓷迹型学系列丛书编辑委员会

**编委会主任：**
胡国华（广东省委宣传部原常务副部长）

**编委会副主任：**
何穗鸿（广州市番禺区文广新局局长）
冯松林（中国科学院高能物理研究所研究员）
江建新（景德镇陶瓷考古研究所副所长、研究员）
肖洽龙（广东省文物鉴定站站长、研究员）
刘成基（广东省考古研究所副所长、研究员）

**总顾问：**
徐尚武（第十届广东省政协副主席）

**顾问：**
齐晓光　许建林　李　穗　苏慧勤

**编委：**
陈少湘　曾昭明　赵国华　周　勇　杨影志
尹春雷　张海斌　肖　华　刘志光　周筱慧
马士宁　周　猛　周　强　文剑华　许以伟
郑　阳　王　灿

**作者：**
周勇（哲学学士，广州大学艺术系原讲师，现任广州东方博物馆馆长）
周强（化工硕士，高级工程师，曾任职广州市化学工业研究所，现任广州东方博物馆副馆长）

**翻译：**
周筱慧　马士宁（Matthew Bunney）　戴灵（Debby Bunney）

# The Ceramics Trace Model Study Series Editing Committee

**Chief Editor:**

Hu Guohua, Guangdong Provincial Party Committee Propaganda Department, Assistant Minister of Operations

**Vice Editors:**

He Suihong, Guangzhou Panyu District Culture, Broadcast, and News Department, Department Head

Feng Songlin, Institute of High Energy Physics – Chinese Academy of Science, Research Fellow

Jiang Jianxin, Jingdezhen Ceramics Archeology Research Institute, Vice Director and Research Fellow

Xiao Qialong, Guangdong Cultural Relic Authentication Station, Station Director and Research Fellow

Liu Chengji, Guangdong Archeology Research Department, Vice Director and Research Fellow

**Chief Consultants:**

Xu Shangwu, 10th CPPCC Guangdong Vice Chairman

**Consultants:**

Qi Shaoguang, Xu Jianlin, Li Sui, Su Huiqin

**Editing Staff:**

Chen Shaoxiang, Zeng Zhaoming, Zhao Guohua, Zhou Yong, Yang Yingzhi, Yin Chunlei, Zhang Haibin, Xiao Hua, Liu Zhiguang, Annabelle Bunney, Matthew Bunney, Zhou Meng, Zhou Qiang, Wen Jianhua, Xu Yiwei, Zheng Yang, Wang Can

**Authors:**

Zhou Yong, B.A. Philosophy, Former Professor of Aesthetics Guangzhou University, Current Director of Guangzhou Oriental Museum

Zhou Qiang, MA.Chemical Engineering, Senior Engineer, Former Research Fellow at Guangzhou Chemical Industry Research Institute, Current Vice Director of Guangzhou Oriental Museum

**Translation:**

Annabelle Bunney (Zhou Xiaohui), Matthew Bunney, Debby Bunney

# 科研项目牵头单位与合作单位
## Sponsors and Partners

**牵头单位：**

广州番禺博物馆

广州东方博物馆

广东省收藏家协会

东方元青花迹型物证研究中心

**合作单位：**

景德镇陶瓷考古研究所

中国科学院高能物理研究所

广东省文物鉴定站

内蒙古包头博物馆

内蒙古明博草原文化博物馆

鄂尔多斯青铜器博物馆

鄂尔多斯农耕游牧博物馆

台湾鸿禧美术馆

泰国曼谷大学东南亚陶瓷博物馆

美国洁蕊堂

美国波斯顿艺术博物馆

美国旧金山亚洲艺术博物馆

美国德克萨斯州金贝尔艺术博物馆

广州大学美术与设计学院

华南师范大学公共管理学院

**Primary Research Project Sponsors**

Guangzhou Panyu Museum

Guangzhou Oriental Museum

Guangdong Collectors Association

The Oriental Material Evidence Research Center on Yuan Dynasty Underglaze Blue Porcelain

**Research Project Partners**

Jingdezhen Ceramics Archeology Research Institute

Institute of High Energy Physics – Chinese Academy of Science

Guangdong Cultural Relics Authentication Station

Inner Mongolia Baotou Museum

Inner Mongolia Mingbo Grasslands Culture Museum

Inner Mongolia Ordos Bronze Ware Museum

Inner Mongolia Ordos Farming and Nomadic Museum

The Chang Foundation

Southeast Asian Ceramics Museum, Bangkok University

The Stamen Collection

Museum of Fine Arts, Boston

Asian Art Museum of San Francisco

Kimbell Art Museum, Fort Worth

Guangzhou University, School of Art and Design

South China Normal University, School of Public Administration

# 番禺博物馆简介
## Guangzhou Panyu Museum

　　番禺博物馆坐落于广州市番禺区银平路龟岗东麓，于1997年11月落成开放，是国家二级博物馆，番禺八大美景之一。馆区所占面积240亩，分为多功能陈列大楼、番禺文博园、东汉古墓群景区等三部分。博物馆展览面积8000平方米，内容丰富，形式多样，表现出鲜明的地方特色；番禺文博园占地面积约30亩，古民居、石桥、牌坊、里弄、古炮、传统作坊等遗迹与田园水乡融于一体，景观宜人。墓葬景区规模宏大，并进行原貌展示，独具特色。

　　番禺博物馆是国家文物局首批展示与服务提升项目的试点单位之一。一楼设"冼星海纪念馆"；二楼设有"番禺古代文明"的3个展厅和多功能临展展厅；三楼设有"馆藏文物"、"名家书画"、"尹积昌雕塑艺术"等多个展厅，陈列独具匠心，融思想性、知识性、艺术性为一体，充分地展示了番禺灿烂的古代文明和今天的辉煌成就。

Panyu Museum is located in Guangzhou City, Panyu District along the Eastern edge of the Turtle Hill Reserve. Established in November 1997, it is a National Level 2 Museum and one of the '8 Scenic Locales of Panyu'. Sitting on 40 acres, the complex is divided into 3 main sections: The Main Multi-Purpose Building, Cultural Garden, and Han Dynasty Burial Grounds. The Main Building consists of 8000 square meters of Display Area with a multifaceted and abundant array of cultural relics from the local area. The Cultural Garden consists of 4 acres and includes: Ancient Peasant Buildings, Stone Bridge, Relic Cannons, and traditional Chinese garden fixtures.

Panyu Museum is one of the first National Cultural Bureau's experimental projects for an elevated level of Display and Service. The First floor is dedicated for the Memorial Hall for Musical Composer Xian Xinghai, a Chinese hero and musical legend. The second floor is divided in to three 'Panyu Ancient Culture' exhibitions as well as one multi-purpose hall. The third floor is allocated to housing Museum Relics, Special Exhibitions, and Yi Zhichang Sculpture Hall. This Museum is dedicated to showing the wonderful ancient culture of Panyu as well as the modern accomplishments that this fine area boasts.

# 广州东方博物馆简介
## Guangzhou Oriental Museum

广州东方博物馆，占地30亩，总建筑面积约8000平方米，由中国建筑设计研究院设计，建筑群体简而不陋、素而不苍，姿横掉阖、藏风纳气。

博物馆按主题展览、专题展览和临时展览之形式分设七个展厅：钟表铜器厅、百砚厅、艺术品交流厅、两个书画厅、两个陶瓷展厅以及四个展廊。

本馆未来的发展方向：科学研究、去伪存真，联合国内权威瓷器研究机构和大专院校，深入展开对陶瓷的量化研究，有效保护、传播、传承中华民族的文化薪火！

Guangzhou Oriental Museum, sitting on nearly 5 acres of land, includes a usable construction area of 8000 ㎡. Designed by China Architecture Design and Research Group (CAG), the entire construction is simple but not meek; plain but not dull; and appears open and expansive but still catches the wind.

The museum is comprised of main exhibitions, themed exhibitions, and special exhibitions spread across seven halls: Bronze Ware Hall, Hundred Ink Stone Hall, Collectors Exchange, two Calligraphy and Painting Halls, and two Ceramics Halls; these halls are linked by four Special Exhibition Corridors.

Guangzhou Oriental Museum's main developmental direction is anchored on: Scientific study, the elimination of bogus pieces and holding on to true relics, partnering with international ceramics studies organizations and major universities, thoroughly developing the quantification research of ceramics, and creating a ceramics research group to take in hand, protect, and pass on the torch of Chinese cultural study!

# 前言

世界上任何一个民族，都有自己的独特文化。这种独特的文化现象，就是这个民族的灵魂。如果一个民族失去了自己的文化，那么，这个民族就不复存在了。

人类许多文化现象，就是多个民族文化在继承、融合、发展中所表现出的现象。一个民族只有不断地摒弃民族文化中的糟粕，传承民族优秀文化，吸收外来优秀文化，才能在世界民族之林中，彰显民族的生存价值和意义，并赢得世界的尊重。

民族的文化，同时也是世界文化的一部分，它当属人类共同文化、共同遗产。各民族文化异而不同，和谐共生。人类只有在彼此认同，彼此尊重不同民族的文化时，世界才会和谐发展，人类才能和平共处。

瓷器是中华民族的发明与创造，研究瓷器文化是为了更好地传承瓷器文化，弘扬瓷器文化。瓷器文化，是中华民族文明进程中重要的文化现象，它已经是世界文化的一部分，属于人类的共同文化遗产。这一点，我们在元青花研究的过程中感受尤深。

伊朗国家博物馆是全世界元青花收藏最多的博物馆，当受邀到伊朗国家博物馆研究元青花时，广州东方博物馆美籍研究人员Mr.Bunney怕遇到签证麻烦，伊朗国家博物馆馆长Asadollah Mohammadpour说："没问题，我来帮你办签证。"可见瓷器文化研究已超越了信仰、国界、政治、种族，因为它属于全人类。

而美国波士顿美术博物馆的人员对我们十分友好，他们尽心为我们的科研需要服务，尽量去满足我们的科研需要。他们对瓷器文化研究，不分彼此，无论国界，境界之高，让人感慨。

在泰国东南亚陶瓷博物馆做研究时，馆长Pariwat Thammapreechakorn亲自下厨款待我们，令人倍感宾至如归。

文物收藏、文物研究本来的要义是解读器物所承载的历史文化信息，享受文化遗存的物件所带来的审美愉悦。如今，因为瓷器收藏已经普及到大众，只有鉴别出它们的真假，才能有效解读历史文化信息，真正享受到文物所带来的审美愉悦。如果收藏不分真假，如果只关注文物的经济效益，而不注重提高真假识别能力和文化研究水平，除了经济损失，最终更将挫伤收藏者的文化自信、文化爱心。

如何让更多的人识真断假，将文物真品、文物精品保护传承下去，就是陶瓷迹型科研项目要解决的主要问题。

陶瓷迹型研究属于物证研究。在陶瓷迹型研究的过程中，从来就不缺乏陶瓷迹型的客观物证，缺的是发现迹型物证的眼光。作为一门新学科，创建的过程，艰难而枯燥，但执着的追求和坚定的付出，让我们终有成效。

在科学研究过程中，我们得到了国内外众多博物馆、研究机构的支持与合作，获得了众多专家、学者、友人的帮助和参与，在本书出版之际，一并衷心感谢。

广州番禺博物馆馆长　曾昭明
广州东方博物馆馆长　周　勇
广东省收藏家协会主席　陈少湘

# Foreword

Regardless of which nation on earth, all have their own distinct culture; it is the soul of the nation. If this culture is somehow lost, this nation ceases to exist.

All of humankind's various cultures have come about through a phenomenon of inheritance, merging, and development of each nation's culture. It is only through the continual cultural distillation (passing on the nation's beneficial culture, and absorbing benefits from foreign cultures) that a nation can stand out among the forest of other world cultures. Only then can they show the world the value and meaning of their existence, and earn the respect of other nations.

National culture, at the same time, is also just a part of a conglomerated World culture, which is the harmonization of every nation's different cultures. Only as we recognize differences and esteem other national cultures, can we, as humankind, converge towards harmony and peaceful coexistence.

Porcelain was the invention of the Chinese people, and through this research we hope to better pass down and even enhance porcelain culture. Even though porcelain culture is an integral facet in the progression of Chinese civilization, it is already considered more than that; a jointly shared product of mankind, a part of World culture. It was through our research of Yuan Dynasty Underglaze Blue (Blue and White) Porcelain that this fact resonated deeply.

The National Museum of Iran houses arguably one of the most important collections of Yuan Underglaze Blue Porcelain. When Matthew Bunney, a member of our research team from the United States, received an invitation from the National Museum of Iran to go and conduct research, he was apprehensive about the trouble in acquiring a visa for the visit. However, Mr. Asadollah Mohammadpour, the museum's director, assured him: "It's no problem, I will help you get the visa." Through this experience, we could see that porcelain cultural research belongs to all mankind, and is already going beyond religion and borders, extending outside politics and nationality.

The Museum of Fine Arts in Boston, offered us their entire collection for scientific research. When greeted by a staff that was accommodating and friendly, which went above and beyond simple hospitality to make us feel welcome, we deeply felt their heart for furthering the research of porcelain culture.

At Bangkok University's Southeast Asian Ceramics Museum in Thailand, the museum director, Mr. Pariwat Thammapreechakron, displayed amazing kindness by not only receiving us but also offering his own cooking services to give us a heartfelt welcome that we will hold with us.

Originally the priority of cultural relic collection and research was to explain and pass on the cultural information offered by a specific piece in order to enjoy the aesthetics of that culture that were carried through time by that ware. However now, as ceramic collection has already entered in to the realm of the general public, if priority is not shifted towards the authentication of such pieces, then the true beauty and information of past cultures will be lost. If collection is done without authentication, only focusing on economic gain rather than placing value on validation and cultural research; then not only will there be economic losses, but the collectors' trust and love of culture will be damaged.

How to allow more individuals to enter in to authentication and thus encourage the preservation of more genuine and fine wares is the one of the main problems that Ceramics Trace Model Study seeks to resolve.

As this is in fact a type of material evidence research, there has never been a lack of the objective material evidence that ceramic traces offer; only a lack of the collective insights to discover them.

The emergence of any new field of study, although a trailblazing scientific work, is arduous and at times tedious. After several years of perseverance, outcomes have finally come to fruition.

During the course of scientific research, we received the support and cooperation of museums and research institutions both domestically and abroad. Additionally, we received much participation and help from experts, academics, and friends. The research of porcelain culture truly belongs to everyone.

With the publication of this book, we would like to take time to thank all the organizations, colleagues and friends that have followed, supported and participated in this research.

Guangzhou Panyu Museum Director: Zeng Zhaoming
Guangzhou Oriental Museum Director: Zhou Yong
Guangdong Collectors Association Chairman: Chen Shaoxiang

# Translation Note

Translation from any language into another is quite a formidable task, as nuances and expressions are often simply unable to be translated across languages. Translations done word for word often carry a certain rigidity that hinders the readability of the text, but a rough transliteration can lead to an understatement of the original meaning. Thus, very different translations of the same text can arise; for example: the Bible has roughly 900 different English translations and paraphrases alone.

As this is a very technical text, this is a relatively literal translation of the original Chinese; while still trying to retain a certain amount of interest and readability. There are certain expressions or idioms that were used that have been converted or paralleled into English in a way that should be more understandable for the reader, but still retain the general meaning of what is trying to be expressed. Throughout the text, conformity to international translating standards was adhered to as much as possible, but in some instances an internationally accepted or consistent version of a certain term or word has not yet been established, so translator discretion was used. As an example: The simplified term 元青花 (yuan qing hua) is the Chinese term for Yuan Dynasty Underglaze Blue Porcelain, which has taken on several other names in the West including Yuan Blue and White, Under the Glaze Cobalt-blue, and so forth. This text utilizes the former and often simplifies this as Yuan Underglaze Blue, or just simply YUB.

Additionally, there were also terms or expressions used that have no specific English translation, or the English is simply the pinyin (or Romanized) version of the Chinese word or phrase. These include certain materials or colors used within the Chinese ceramics vocabulary. As an example: 青 qing, which is a color that can take on a whole variety of forms from green, to blue, to clear, and so forth has even offered confusion for Chinese writers throughout history (we will discuss this more within the text). Also, commonly interchangeable terms such as: Silica, Silica Glass, Silicate, Silicone Dioxide, and so forth can all refer to the material with the chemical representation $SiO_2$, and are similarly used as such according to original context.

In the case of titles of books which do not have the title or text translated into English, the original pinyin is used for reference and ease of use in catalogue recall. Quotes from Chinese texts and sources have been independently translated for this publication, and any other translations that currently do exist, if any, were not consulted. In the case that a quotation in the Chinese text was translated originally from an English source, then the original text was used in the English translation and cited as such. Chinese authors' names utilize Chinese pinyin citation; whereas non-Chinese names retain Western spelling.

Additionally, due to the relatively trail-blazing nature of this research, many new terms and vocabulary were literally invented for the purpose of this survey and further research. Specifically, the term 'Ceramics Trace Model Study' had to be invented in both Chinese and English in order to capture the essence of this groundbreaking research. Fortunately, many of the new terms are names and descriptions of appearances and characteristics encountered throughout the use of Ceramics Trace Model Study. As such, English translations have followed as closely to the original Chinese as possible, straying only in an attempt to more vividly represent said characteristics. In the case of newly introduced Chinese terms, all care was taken to retain consistency in translation. However, any discrepancy should be conceded to the original Chinese text.

Similar to the nature of Ceramics Trace Model Study, this translation is also open and ongoing. We welcome academics, researchers, and other readers to actively participate in the continual improvement of its content for the betterment of human knowledge! Thank you for your understanding and continued involvement in our research.

Guangzhou Oriental Museum Research Fellow: Zhou Xiaohui
Guangzhou Oriental Museum Deputy Director: Matthew M. Bunney

# 目 录

## 第一章　陶瓷迹型学产生背景 / 2

**第一节**　社会需求催生陶瓷迹型学 / 4
**第二节**　陶瓷观察工具发明与使用 / 6
　　一、陶瓷观察工具发明 / 6
　　二、陶瓷观察工具使用 / 6
**第三节**　陶瓷迹型学产生的政治、文化背景 / 10
　　一、传统体制原因 / 10
　　二、现实体制原因 / 10
　　三、传统文化原因 / 12
　　四、传统思维原因 / 12
　　五、传统鉴定方法存在局限性 / 12
　　六、入门要求原因 / 14
　　七、小结 / 14
**第四节**　陶瓷鉴定方法的发展演变 / 18
　　一、口说为凭经验证据 / 18
　　二、客观实在物证证据 / 18
　　三、物证系统——陶瓷迹型学创立 / 22
　　四、小结 / 24

## 第二章　陶瓷迹型学的宗旨、方法与意义 / 28

**第一节**　陶瓷迹型学研究宗旨与方法 / 30
　　一、陶瓷迹型学研究宗旨 / 30
　　二、陶瓷迹型学研究方法 / 30
**第二节**　陶瓷迹型学研究意义 / 36
　　一、建构陶瓷迹型标准图谱模型 / 36
　　二、标准图谱模型比对鉴定易于大众掌握 / 36
　　三、从听鉴定结论到看鉴定结论 / 38
　　四、从鉴定人证到鉴定物证 / 38
　　五、陶瓷迹型学是陶瓷鉴定方法的
　　　　一种补充 / 38
　　六、陶瓷迹型学是历史发展的必然趋势 / 40
　　七、陶瓷迹型学是人类认识发展的必然 / 40

## 第三章　陶瓷迹型学 / 46

**第一节**　陶瓷迹型学基础概念 / 48
　　一、陶瓷迹型学概念系列一 / 48
　　二、陶瓷迹型学概念系列二 / 50
**第二节**　传统陶瓷鉴定依据 / 58
　　一、传统陶瓷鉴定依据 / 58
　　二、关于类型学 / 58
　　三、传统鉴定依据的局限性 / 60
**第三节**　陶瓷迹型学鉴定依据 / 64
　　一、陶瓷迹型及规律是鉴定的主要依据 / 64
　　二、陶瓷迹型标准图谱模型是鉴定的
　　　　物证系统 / 64
**第四节**　陶瓷迹型学鉴定原理 / 68
　　一、陶瓷迹型现象存在的必然性原理 / 68
　　二、陶瓷迹型量变原理 / 68
　　三、陶瓷迹型变化时间过程原理 / 68
**第五节**　陶瓷迹型学鉴定本质 / 74
　　一、观察整体化 / 74
　　二、比对证据客观化 / 74
　　三、论证逻辑化 / 76
　　四、求证实验化 / 76
**第六节**　陶瓷迹型学局限性 / 80
　　一、标本选择的局限性 / 80
　　二、显微研究的局限性 / 80

三、人工操作的局限性 / 80
　　四、陶瓷迹型缺乏的局限性 / 82

## 第四章　陶瓷釉层的化学成分、性质与形貌 / 84

第一节　陶瓷釉化学成分 / 86
　　一、釉的概念 / 86
　　二、釉的构成 / 86
　　三、釉的分类 / 86
　　四、釉的配方 / 92
第二节　陶瓷釉化学成分的作用 / 94
　　一、网络形成剂作用 / 94
　　二、助熔剂作用 / 94
　　三、其他成分作用 / 98
　　四、釉的化学成分与釉层结构稳定性的关系 / 98
第三节　陶瓷釉表面活性 / 100
　　一、釉表面活性概念 / 100
　　二、釉表面活性产生原因 / 100
　　三、小结 / 100
第四节　陶瓷釉层形貌 / 102
　　一、釉层概念 / 102
　　二、釉层形貌 / 102
第五节　总结 / 110

## 第五章　陶瓷釉层变化 / 112

第一节　陶瓷釉层变化机理 / 114
　　一、釉层与水的反应 / 114
　　二、釉层与碱的反应 / 116
　　三、釉层与酸的反应 / 116
　　四、釉层与$CO_2$的反应 / 118
　　五、釉层与盐溶液的反应 / 118
　　六、釉层风化产物 / 118
　　七、小结 / 118
第二节　陶瓷釉层变化与环境关系 / 124
　　一、地上环境 / 124
　　二、地下环境 / 124
　　三、水中环境 / 126

第三节　釉层迹型风化特征 / 130
　　一、玻璃相迹型风化特征 / 130
　　二、气相迹型风化特征 / 130
　　三、晶相迹型风化特征 / 130
第四节　总结 / 134

## 第六章　陶瓷典型迹型类别 / 136

第一节　元青花斑块类别 / 138
　　一、元青花斑块概念 / 138
　　二、元青花斑块类型 / 138
第二节　元青花纹路类别 / 168
　　一、元青花纹路概念 / 168
　　二、元青花纹路种类 / 168
第三节　元青花气泡类别 / 182
　　一、元青花气泡概念 / 182
　　二、元青花气泡种类 / 182
第四节　总结 / 206

## 第七章　陶瓷迹型规律 / 208

第一节　陶瓷迹型随机性规律 / 210
　　一、陶瓷迹型随机性规律概念 / 210
　　二、陶瓷迹型随机性规律成因 / 210
　　三、陶瓷迹型随机性规律表现特征 / 210
　　四、小结 / 212
第二节　陶瓷迹型色差规律 / 214
　　一、陶瓷迹型色差规律概念 / 214
　　二、陶瓷迹型色差规律成因 / 214
　　三、陶瓷迹型色差规律表现特征 / 214
　　四、小结 / 214
第三节　陶瓷迹型量变规律 / 218
　　一、陶瓷迹型量变规律概念 / 218
　　二、陶瓷迹型量变规律成因 / 218
　　三、陶瓷迹型量变规律表现特征 / 218
　　四、小结 / 222
第四节　陶瓷迹型层次性规律 / 224
　　一、陶瓷迹型层次性规律概念 / 224
　　二、陶瓷迹型层次性规律成因 / 224

三、陶瓷迹型层次性规律特征 / 224
　　　四、小结 / 226
第五节　陶瓷迹型多样性规律 / 228
　　　一、陶瓷迹型多样性规律概念 / 228
　　　二、陶瓷迹型多样性规律成因 / 228
　　　三、陶瓷迹型多样性规律特征 / 228
　　　四、小结 / 228
第六节　陶瓷迹型差异性规律 / 230
　　　一、陶瓷迹型差异性规律概念 / 230
　　　二、陶瓷迹型差异性规律成因 / 230
　　　三、陶瓷迹型差异性规律特征 / 230
　　　四、小结 / 230
第七节　陶瓷迹型分明性规律 / 232
　　　一、陶瓷迹型分明性规律概念 / 232
　　　二、陶瓷迹型分明性规律成因 / 232
　　　三、陶瓷迹型分明性规律特征 / 232
　　　四、小结 / 232
第八节　总结 / 234

附录 / 236

参考文献 / 256

# CONTENTS

**Chapter 1: The Emergence of Ceramics Trace Model Study / 3**

Section 1: Society's Need to Induce Ceramics Trace Model Study / 5

Section 2: The Invention and Use of Ceramic Observation Tools / 7
1. Invention of Ceramic Observation Tools / 7
2. Use of Ceramic Observation Tools / 7

Section 3: The Political and Cultural Background of the Emergence of Ceramics Trace Model Study / 11
1. Traditional System Influence / 11
2. Actual System Influence / 11
3. Traditional Culture Influence / 13
4. Traditional Ideology Influence / 15
5. Traditional Authentication Method Has Limitations / 15
6. Threshold Influence / 17
7. Conclusion / 17

Section 4: Evolution of Ceramic Authentication Methods / 19
1. Oral Account Conveying Experiential Evidence / 19
2. Objective Material Evidence / 21
3. Material Evidence System – Established by Ceramics Trace Model Study / 25
4. Summary / 27

**Chapter 2: The Purpose, Method, and Significance of Ceramics Trace Model Study / 29**

Section 1: The Purpose and Method of Ceramics Trace Model Study / 31
1. The Purpose of Ceramics Trace Model Study / 31
2. The Method of Ceramics Trace Model Study / 31

Section 2: The Significance of Ceramics Trace Model Study / 37
1. Construct Ceramic Standard Trace Atlas Model / 37
2. Atlas Model Comparison Easier for Layman to Grasp / 37
3. From Hearing Appraisal to Seeing Authentication / 39
4. From Human Witness Appraisal to Material Evidence Authentication / 41
5. Trace Model Study Complementary to Other Authentication Methods / 41
6. Historical Development's Necessity for Ceramics Trace Model Study / 43
7. Human Knowledge Development's Necessity for Ceramics Trace Model Study / 44

**Chapter 3: Ceramics Trace Model Study / 47**

Section 1: Ceramics Trace Model Study Foundational Concepts / 49
1. Ceramics Trace Model Study Concepts Series 1 / 49
2. Ceramics Trace Model Study Concepts Series 2 / 51

## Section 2: The Foundations of Traditional Ceramic Appraisal / 59

1. The Foundations of Traditional Ceramic Appraisal / 59
2. Typology / 61
3. Limitations in the Foundation of Traditional Appraisal / 61

## Section 3: The Foundations of Ceramics Trace Model Study / 65

1. Ceramic Traces and Ceramic Trace Laws as the Primary Foundation of Authentication / 65
2. Standard Trace Atlas Model as the Evidence System for Authentication / 65

## Section 4: The Principles of Ceramics Trace Model Study / 69

1. The Inevitability Principle / 69
2. The Quantitative Change Principle / 69
3. The Passage of Time Principle / 69

## Section 5: The Nature of Ceramics Trace Model Study / 75

1. Holistic Observation / 75
2. Objective Evidence / 75
3. Logical Arguments / 77
4. Experimental Verification / 77

## Section 6: Limitations of Ceramics Trace Model Study / 81

1. Limitations of Sample Selection / 81
2. Limitations of Microscopic Research / 81
3. Limitations of Manual Operation / 83
4. Limitations of Lacking Traces / 83

# Chapter 4: Chemical Composition, Characteristics and Appearance of the Ceramic Glaze Level / 85

## Section 1: Chemical Composition of Ceramic Glaze / 87

1. The Definition of Glaze / 87
2. The Components of Glaze / 87
3. Types of Glaze / 87
4. Formulation of Glaze / 91

## Section 2: The Functions of Ceramic Glaze Chemical Components / 95

1. The Function of Networking Agents / 95
2. The Function of Flux / 95
3. The Function of Other Components / 97
4. The Relationship Between Glaze Composition and Glaze Stability / 97

## Section 3: Active Nature of the Ceramic Glaze Surface / 101

1. The Active Nature of the Glaze Surface Defined / 101
2. Causes of the Glaze Surface Activity / 101
3. Summary / 101

## Section 4: Ceramic Glaze Level Appearance / 103

1. Glaze Level Defined / 103
2. Glaze Level Appearance / 103

## Section 5: Conclusion / 111

# Chapter 5: Ceramic Glaze Level Changes / 113

## Section 1: Mechanics of Glaze Level Change / 115

1. Glaze Level Reactions with Water / 115
2. Glaze Level Reactions with Alkali / 117
3. Glaze Level Reactions with Acid / 119
4. Glaze Level Reactions with Carbon Dioxide / 119
5. Glaze Level Reactions with Salt Solutions / 121
6. Sedentary Byproducts / 121
7. Summary / 122

## Section 2: Relationship Between Environment and Glaze Level Change / 125

1. Above Ground Environment / 125
2. Underground Environment / 125
3. Underwater Environment / 127

## Section 3: Characteristics Of Weathered Glaze Level Traces / 131

1. Characteristics of Weathered Glass Phase Traces / 131

2. Characteristics Of Weathered Gas Phase Traces / 133

3. Characteristics Of Weathered Crystallized Traces / 133

Section 4: Conclusion / 135

## Chapter 6: Typical Ceramic Trace Categories / 137

Section 1: The Yuan Dynasty Underglaze Blue Porcelain Mottle Category / 139

1. YUB Mottles Defined / 139

2. YUB Mottle Classes / 139

Section 2: The Yuan Dynasty Underglaze Blue Porcelain Striae Category / 169

1. YUB Striae Defined / 169

2. YUB Striae Types / 169

Section 3: The Yuan Dynasty Underglaze Blue Porcelain Air Bubble Category / 183

1. YUB Air Bubble Defined / 183

2. YUB Air Bubble Types / 183

Section 4: Summary / 207

## Chapter 7: Ceramic Trace Laws / 209

Section 1: Ceramic Trace Law of Randomness / 211

1. Ceramic Trace Law of Randomness Defined / 211

2. Ceramic Trace Law of Randomness Causes / 211

3. Ceramic Trace Law of Randomness Characteristics / 211

4. Summary / 213

Section 2: Ceramic Trace Law of Chromatic Aberration / 215

1. Ceramic Trace Law of Chromatic Aberration Defined / 215

2. Ceramic Trace Law of Chromatic Aberration Causes / 215

3. Ceramic Trace Law of Chromatic Aberration Characteristics / 215

4. Summary / 215

Section 3: Ceramic Trace Law of Quantitative Change / 219

1. Ceramic Trace Law of Quantitative Change Defined / 219

2. Ceramic Trace Law of Quantitative Change Cause / 219

3. Ceramic Trace Law of Quantitative Change Characteristics / 219

4. Summary / 221

Section 4: Ceramic Trace Law of Gradation / 225

1. Ceramic Trace Law of Gradation Defined / 225

2. Ceramic Trace Law of Gradation Causes / 225

3. Ceramic Trace Law of Gradation Characteristics / 225

4. Summary / 227

Section 5: Ceramic Trace Law of Variegation / 229

1. Ceramic Trace Law of Variegation Defined / 229

2. Ceramic Trace Law of Variegation Cause / 229

3. Ceramic Trace Law of Variegation Characteristics / 229

4. Summary / 229

Section 6: Ceramic Trace Law of Disparity / 231

1. Ceramic Trace Law of Disparity Defined / 231

2. Ceramic Trace Law of Disparity Cause / 231

3. Ceramic Trace Law of Disparity Characteristics / 231

4. Summary / 231

Section 7: Ceramic Trace Law of Distinction / 233

1. Ceramic Trace Law of Distinction Defined / 233

2. Ceramic Trace Law of Distinction Causes / 233

3. Ceramic Trace Law of Distinction Characteristics / 233

4. Summary / 233

Section 8: Conclusion / 235

## Appendix / 236

## References / 256

# 第一章
# 陶瓷迹型学产生背景

**阅读提示：**

　　任何一门学科的诞生，都不是一帆风顺的。但是，社会发展、科学发展、民众需要，催生新学科诞生，又是无法阻挡的。社会发展史和科学发展史都说明了这一事实。本章重点讨论陶瓷迹型学产生的社会、历史、文化等背景原因，并阐明陶瓷鉴定方法的演变以及陶瓷迹型学产生的必然性。

**主要论述问题：**

　　1. 为什么显微镜从问世到运用于中国陶瓷显微研究相隔了400多年？
　　2. 陶瓷鉴定方法是如何演变发展的？
　　3. 陶瓷迹型学都有哪些创新？

## Chapter 1
# The Emergence of Ceramics Trace Model Study

Text Note:

    The emergence of any branch of learning is brought forth with much labor. Each field is induced through the advancement of society and to meet the needs of the people, and there is often no stopping it. Throughout history, scientific and societal developments have all pointed to this. Therefore, this chapter discusses not only the social, historical, cultural, and other implicated difficulties in the emergence of Ceramics Trace Model Study, but also the necessity of the birth of this field in the context of the progression of ceramics authentication methods.

## 第一节
# 社会需求催生陶瓷迹型学

任何一门学科的诞生，都是人类智慧的结晶，都是社会和民众的巨大需求推动的结果。

改革开放30年来，中国发生了巨大变化。中国人从生存状态走向了生活状态，走向了品质生活状态、精致生活状态。不同的生活状态，就有不同的精神生活追求。富起来的中国人越来越多地把目光转向了文化艺术品收藏。但是，如今中国的收藏市场，乱象纷飞，尤其是陶瓷收藏市场，尔虞我诈，真假难辨。其中一个重要原因在于：目前中国的陶瓷鉴定依然停留在人证为主的经验性时代。

人证为主的鉴定时代，没有量化的、科学的客观标准，不像钻石鉴定已经步入了物证为主的科技时代，有量化的科学等级标准。在中国的陶瓷鉴定领域中，由于缺乏公正的裁判规则及客观物证，陶瓷鉴定界的公信力遭到公众挑战。

胡适先生曾经说过："一个民族和族群，总是在提倡道德而不注重规则的话，这个民族会道德沦丧的。"所以，如果陶瓷鉴定没有科学的标准，久而久之，陶瓷市场就会假货充斥，社会伦理、社会诚信的底线就会崩溃，延续千年的文脉就会断裂，社会管理将会为此付出巨大成本。

这将会挫伤收藏者和投资者的信心，会毁掉中国新兴繁荣的文物市场，还会挫伤中国人民保护及传承民族文化的爱心，使中国人民丧失对民族文化的自信。

所以，民众、市场乃至国家的诚信，社会的稳定，都迫切需要陶瓷鉴定界从人证的经验时代走向物证的科技时代，走向物证的客观公正时代，走向有客观评判规则的时代。

社会需求、民众基础、国家文化政策支持、科学技术允许等，这些条件促成了陶瓷迹型学的诞生。其中热爱中华民族文化、热爱收藏的广大民众的迫切需求，正是陶瓷迹型学产生的重要原因。这一点与其他学科不同，有些学科的诞生，往往要去诱导和唤醒民众的需求。

# Section 1:
# Society's Need to Induce Ceramics Trace Model Study

Every branch of learning is the crystallization of man's wisdom; the born fruits of society's needs.

Since opening to the world over 30 years ago, China has gone through vast changes. Chinese have progressed from a state of survival, to a state of livelihood, to a state of quality, to a state of exquisiteness. Different lifestyles mean a different vigor in lifestyle pursuits. There is an increasing number of wealthy Chinese that are placing their gaze on the collection of cultural works of art. However, today the Chinese art market has become a muddled blizzard; specifically ceramics collection, where a cut-throat environment of being unable to distinguish real from fake has arisen. This is primarily due to the fact that ceramics certification still remains in an era of relying on human experience or witness as the main authority.

An era that relies upon human witness as the main authority lacks clear-cut precision and is void of quantitative scientific observation standards, unlike those seen in the authentication of precious stones, which has moved into a technological era of material evidence as authoritative proof and uses a quantitative scientific observation standard. The realm of authenticating Chinese ceramics is lacking fair judgment principles and objective material evidence. Therefore, experts' authority has run upon a gauntlet of public distrust and the questioning eye of the market.

Mr. Hu Shi (famous Modern Chinese scholar) once said, "If a nation or an ethnic group always emphasizes morality but ignores rules, it will eventually lose its morality." Ceramics authentication without scientific standards (over time and as more and more fakes flush into the market) causes societal morals and basic honesty to begin to crack and huge repercussions will be felt.

This in turn will hurt the protection and inheritance of Chinese culture and bruise collectors' and investors' faith. Chinese will lose faith in their own culture, and the Chinese's newly developing and flourishing cultural relic market will be destroyed. Most importantly, damage will be done to the love and protection of Chinese culture and heritage.

As such, for the sake of the public, the market, as well as national credibility, integrity, and societal stability there is an urgent need for the ceramics authentication world to move from an era of authority relying on human witness into a technological era in which material evidence provides objective and impartial authority.

Many factors such as society's need, people's need, governmental support through cultural policies, and science and technological capabilities, have facilitated the emergence of Ceramics Trace Model Study. Unlike the emergence of some other branches of learning that have required guidance to awaken societies' needs, Ceramics Trace Model Study has been urgently induced and born out of the requisite from societal development, passion for Chinese Culture, and love of collecting authentic works of art.

# 第二节
# 陶瓷观察工具发明与使用

## 一、陶瓷观察工具发明

1590年，荷兰约翰逊父子发明光学显微镜，一个全新的显微世界展现在人类的视野中。显微镜的发明为现代医学、生物学、植物学、仿生学、化学等诸多学科奠定了基础。

1609年，意大利科学家伽利略发明天文望远镜，从此人类的视野进入了宏观世界。天文望远镜的发明为现代天文学、航天学的发展奠定了基础。

显微镜和望远镜是人类最伟大的发明之一。

科学的发明必然带来科技的进步，从而推动人类文明的进步。科学的发明，科技的进步，总是在延伸人类的器官，使人类的观察认知逐渐突破肉眼观察的局限。人类借助显微镜和望远镜，将视野延伸到了精美的微观世界与广袤的宏观世界中。

借助科学仪器和技术手段，对陶瓷进行显微及微观研究是历史及社会发展的必然进程，也是人类不断深入认识世界的必由之路，因为科学发展史就是一部从对事物定性的模糊研究发展到定量的精确研究的历史。但是，从显微镜的发明到将其应用在陶瓷研究领域，在中国却是400多年之隔。

## 二、陶瓷观察工具使用

### （一）20世纪初的情况

20世纪初，即使有显微镜，也不被用来观察陶瓷。

那时人们观察陶瓷基本是用10倍以下的放大镜。用放大镜观察陶瓷的目的，也只是想看得更清楚一些，并非想要通过陶瓷的显微及微观迹型特征去判断陶瓷的真假。即使能意识到以显微及微观观察的方式去判断陶瓷真伪，但由于观察工具的局限性，也是"有心无力"。

### （二）建国后的情况

1949年到1979年，中国就是一个"折腾"的时代，基本没有文物市场，每个省市只保留极少体制内的文物商店。1966年，"文革"开始，大破"四旧"，否定民族文化，这对中华民族文化可谓一场浩劫。古陶瓷本身能免遭损毁就是万幸了，更别说对古陶瓷迹型的深度研究。那是历史上陶瓷鉴定与研究的黑暗时期，整个国家、整个行业对古陶瓷的研究鉴定"无心无力"。

自1979年改革开放以后，随着综合国力的持续上升，国家对文物市场的逐步开放，中国的收藏队伍日益壮大。他们当中，既有潜心研究的专家学者，更有广大文物的爱好者，还有投资者与捐客。

但直至20世纪90年代，市场上还很难买到18倍的放大镜。那时，18倍的放大镜在上海的市场价格是800元人民币左右，而且在极少地方有卖的，因为在市场上，民众对高倍放大镜的需求量是极低的。

### （三）21世纪的情况

如今的古玩市场，造假猖獗。由于某些专家的鉴定水平跟不上陶瓷造假技术的发展，以及某些专家把不住道德底线，造成人民对专家的公信力产生质疑，人民热切盼望能找到一种更加科学客观的陶瓷鉴定方法。因此，利用显微镜研究陶瓷的队伍不断壮大，这夯实了对陶瓷迹型进行探寻的群众基础。

目前使用显微镜的人多半是普通收藏爱好者，而体制内的专家鲜有使用。

对于研究型专家，陶瓷显微及微观研究与鉴定不是他们的工作范围，不能苛求他们去做分外的事。

但在民间，陶瓷显微研究却进行得如火如荼，势不可当。巨大的市场需求，刺激了生产竞争，激烈的生产竞争，导致了激烈的价格竞争。2000年，市场上很难买到100倍以上的放大镜。2003年以后，市场上100倍以上的显微镜价格在

# Section 2:
# The Invention and Use of Ceramic Observation Tools

## 1. Invention of Ceramic Observation Tools

In Holland 1590, the Johnson father-son duo invented the Optical Microscope and brought a whole new microscopic world into people's visual boundaries. Microscopes established a new foundation for Medicine, Biology, Botany, Bionics, Chemistry, and other branches of Science.

In Italy 1609, scientist Galileo invented the Astronomy Telescope and humankind's visual boundaries extended into the macrocosm. It was on the foundation built by the Astronomy Telescope that Astronomy and Aeronautics took form.

To this day both the Optical Microscope and the Astronomy Telescope remain as some of man's greatest inventions.

Scientific invention brings with it technological advancement, which is the driving force in civilizations' progression. Scientific invention and technological advancement help to stretch man-made apparatuses, and have gradually facilitated extending the limits of the human eye and expanding our knowledge and observable limits. The telescope and microscope have extended human's field of vision to include the expansive breadth of the macrocosm and the exquisite beauty of the microscopic world.

Seeking the help of scientific instruments and technological medium to enter the microscopic and microcosmic world in ceramic studies was an imperative step, just as humans continually deepening their understanding of the world around them is a road that must be traveled. Scientific development is the story of moving from obscurely studying a thing to precisely understanding that object. Unfortunately, from the invention of the microscope to the application of it to study Chinese ceramics, over 400 years has passed.

## 2. Use of Ceramic Observation Tools

### A. Early Twentieth Century

Even had there been observation tools available in the Early Twentieth Century, they were not used for looking at ceramics.

During that time the only tool used for inspecting ceramics was a magnifying glass of less than ten times magnification. It was used in making surface observation clearer, and not to look at the microscopic or microcosmic traces on the porcelain to decide its authenticity. Even though some may have recognized the need to examine ceramic traces to authenticate and further study a piece, due to limitations of the observation tool, it was difficult to effectively examine a piece's microscopic qualities. "There may have been a will, but there was no way".

### B. Peoples Republic Era to Today

From 1949-79 China was in a period of struggle, and generally there was no market for cultural relics, save for government run cultural relic stores in each provincial city. In 1966, the Cultural Revolution started, and in an attempt to cleanse old customs and habits, Chinese Culture was denied and attacked. Not only was there no additional study done on Chinese ceramics; the fact that any relics survived is a miracle in itself. It was during that time that ancient ceramics authentication and study completely ceased, the entire country and industry lost interest in furthering development. "There

400—500元。随后的几年中，同样倍数的显微镜，一路跌价，当年几百元的显微镜，如今在网上30元就可以买到。

显微镜需求量大增，造就了庞大的陶瓷显微研究队伍。

民间收藏者或从事古玩生意的人，他们的研究动力来自这两方面：

一是提高自身鉴定水平，从而在目前由于国家管理不严而假货横飞的古玩市场上减少投资的经济损失。

二是研究成果能取得行业认可，争取话语权。

目前，越来越多的陶瓷收藏爱好者加入到陶瓷显微研究的行列中。民间研究力量的崛起，对陶瓷显微研究将起到巨大的推动作用。尽管目前陶瓷显微研究的成果寥寥无几，但我们相信，不久的将来，会有更多相关的论文及著作问世，因为陶瓷的显微研究已经进入到"有心给力"的阶段。

was no will, and no way".

In 1979 China opened its doors to the outside world. As each aspect of the Nation began to develop and grow, the relics industry reopened, and more and more people have become interested in collection. According to polls, there are now more than 80 million people who express interest in relic collection; among them are scholars, amateur collectors, investors, and brokers.

However, up into the Nineties, it was very difficult to even acquire an 18X magnifying glass. At that time, in Shanghai, an 18X magnifying glass cost about 800 RMB and was very difficult to get a hold of because the market had no specific need for a glass of that magnification.

## C. Twenty First Century

Recently the market for forging ceramics has been running wild. Due to some authentication experts falling behind the forger's technology, as well as some experts losing grasp of their moral standards, the public has brought into question the reliability of the expert's opinions. The market is eager to find a more scientific authentication method. Therefore, using microscopes to study ceramics has become more and more popular, thus paving the way for the exploration of Ceramics Trace Model Study.

Currently in China, the majority of people that are using microscopes to explore ceramics are amateur collectors, whilst government sanctioned experts rarely take part.

For government research experts, ceramic microscopic and microcosmic study is not within the realm of their assigned job description, thus cannot be required of them.

Amongst the people however, microscopic study of ceramics is catching fire, an unstoppable trend. In 2000, it was difficult to find a 100X magnification scope on the market; but after 2003, magnification of 100X and above was attainable for 400-500 RMB. From then on, the same magnification scopes have become more widely available and cheaper to purchase. Today you can purchase a 100X magnification scope for 30 RMB online.

This goes to show that the huge demand in the market led to increased and more efficient production, leading to these price wars. This resulted in a large group of people with the capability of doing their own microscopic observations of ceramics.

Amateur collectors and antique dealers have two main motives for research: The first, being to improve their own authentication abilities, thus reducing financial loss in the rampant forgery market that the government has not the means to curtail. The second, being their research achievements can be recognized within the industry, thus building rapport and credibility within the trade.

Now, more and more amateur collectors are entering into the microscopic research of ceramics. The rise in research power amongst these laypersons is precisely the force that will push ceramics microscopic research to the next level. Although currently, the published results of this field of research are very sparse, we believe that future publications and research results will increase in number and fruitfulness, because we have now entered into an era of "There is a will, and there is a way".

# 第三节
# 陶瓷迹型学产生的政治、文化背景

一项工具的发明，并不意味着它马上就能得到广泛运用，它的运用可能会受到社会政治、经济、文化、习俗等各方面的影响。

## 一、传统体制原因

辛亥革命以前，中国一直是君主专制的中央集权国家。家国同构，皇权至上，君主的意志凌驾于民众意志之上，至明清两代这种君主专权制度已发展到极致。

在这种制度下，国人性格趋向自闭、附庸的特征，同时产生演绎附会、揣摩臆测、察言观色、随机应变、模棱两可等行为方式，少有自主创新意识产生。

18—19世纪，西方工业革命蓬勃开展，那时有很多先进的工业科技产品传入中国，比如汽车、照相机、钟表等，但因为皇族和统治阶层的盲目自大与保守的心态，更多的西方先进技术被阻挡在国门之外，科技的火光熄灭在统治阶层的排外意识中。

19世纪中叶，清朝廷看到西方的机器、枪械、轮船的制造，无一不与天文、算学有关，于是恭亲王奕䜣筹办京师同文馆，开设了不少科学实用技术的学科。但是，这些兴科学之举却遭到了朝廷守旧势力的反对。他们认为尊王攘夷，坚持忠信礼义，就能够"制胜自强"，反对师夷而求新求变。

与西方不同的是，工业革命和技术进步都未给中国带来彻底改变社会的革命性影响，其根本原因就是传统的封建体制没有发生根本性变革。

这种制度和观念对陶瓷鉴定领域的影响是：人们将眼学的手段视为金科玉律，所以，陶瓷鉴定中的眼学的方法一直是以"师傅带徒弟"的模式代代相传。由于每个师傅代代相传的陶瓷鉴定知识，不能够形成统一概念的认识结晶，缺乏对规律的揭示与总结，缺乏理论体系的建构与创新，陶瓷鉴定无法归入学制式教育。

故宫老专家李辉柄在《青花瓷器鉴定》一书中提到，长期以来，瓷器鉴定由于没有进行科学的总结，缺乏理论化，只停留在"只能意会，不能言传"的感性阶段。将瓷器鉴定提高到理论研究是鉴定家们奋斗的目标[1]。

## 二、现实体制原因

新中国成立后，由于体制原因，中国陶瓷的研究人员与鉴定人员大部分是分离的。也就是说，从事考古与文物研究的专家的工作范围与职责是不包括鉴定工作的，因此他们没有国家颁发的鉴定牌照。这种分离是因为他们属于不同的单位造成的。

当然也有极少数专家，既搞研究，也持有鉴定资格的牌照，因为政府部门和事业单位的专家是可以相互调动的。

在中国从事研究的人员部分属在大专院校考古系，部分在省市考古研究所及博物馆中。他们日常的首要工作是搞研究，而不是搞鉴定。他们搞研究的特点是研究文物的"真"。他们研究的器物绝大部分是已经确定了年代的，因为这些器物绝大多数来源于考古发掘。一般来说，从考古发掘的器物上可直接获得断代的证据，因此不需要在发掘出的器物上下更多的鉴定功夫。如果碰到年代不详的古墓，他们会邀请从事鉴定的专家协助对发掘器物进行断代，但不需要判断器物真伪。

这类专家的工作及生活经费来源都是由国家提供。出于行业自律，他们不得擅自从事文物交易。因此，他们缺乏来自市场买卖判断真假器物的经验。他们的工作精力主要放在研究"真"的器物上，可以置"假"而不顾。对于进入显微及微观世界探寻鉴定器物的新方法，不是他们工作的主要范围。由于这类专家工作性质的关系，对于他们个人的鉴定素质和能力，组织及单位没有刚性考核要求。

于是，这类专家对工作分外的

Section 3:

# The Political and Cultural Background of the Emergence of Ceramics Trace Model Study

The mere invention of a tool does not mean it will immediately be implemented every place that it is needed; there are additional political, economic, cultural, and conventional factors that act as barriers to its widespread use.

## 1. Traditional System Influence

Before the revolution of 1911, China had always been an autocratic monarchy with centralized governance. Family and National power structure was the same, the power of the monarch supreme, the will of the court superseded the will of the people. During the Ming and Qing Dynasties this monarch power monopoly system reached its most developed state.

Under this system, people have a tendency to keep to their own, keep their heads down, develop controlled mannerisms, bend to the necessity of the situation, and equivocate ambiguity. This greatly stifles creativity and the emergence of new ideologies.

The Eighteenth and Nineteenth Centuries saw the Industrial Revolution in the West. With it, brought the introduction of many new industrial products into China. For example: Vehicles, cameras, clocks, and so forth. However, due to the Imperial and Political Classes' blind arrogance and conservative mentality, additional advanced technology from the West was kept outside the nation's doors. Thus, the flame for new technology was extinguished by the antiforeign ideologies of the ruling class.

In the mid-Nineteenth Century, the Qing Dynasty court began to see that the making of machines, guns, and ships were all related to astronomy and mathematics. Thus, Gong Qin Wang (An important Imperial family member in the Qing court) started relevant schools. Unfortunately, the discoveries of these popular sciences were met by the conservative influences of the Imperial government. They were a conservative communal that honored the Emperor and shunned that which was foreign. Holding that loyalty and integrity was enough to bring greatness, thus they turned away foreign teachings and new changes.

Where their industrialization and technological advancements differed from the West was that they did not bring China thorough society-altering revolutionary impact. This was due large in part to the fact that there was no substantial change to the autocratic monarchy.

The influence that this system and ideology had on ceramic authentication is: People took the examination of ceramics by the naked eye as the gold standard of authentication. Therefore, using the naked eye for examining a piece has always been used and passed down through a deeply ingrained master-apprentice model. As Li Huibing from the Beijing Palace Museum stated, "Porcelain authentication has always been considered as an inexplicable intuition that can only be gained through experience." For a long time, due to the lack of theorizing and technological conclusions, ceramic authentication has remained in a period of "You can only get an idea, but not fully explain". The focus of the experts' struggles is to elevate authentication to theorizing research [1].

## 2. Actual System Influence

Since the establishment of the People's Republic of China, due to the systems in place, the research and authentication of ceramics has, for the most part, remained as two separate fields. This is to say, that the archeological and cultural relic experts' scope of work and responsibility does not include authentication and thus they do not have government recognized licenses to authenticate. This

事情缺乏研究动力。他们在工作中不会因未对陶瓷进行进一步的显微及微观研究而有紧迫感或压力。如果他们过于热衷钻研器物的鉴定，反而容易遭到不务正业的误解。现实中，只有极少部分研究人员开始关注文物鉴定并参与到其中。当社会需要学者型研究人员去做鉴定的时候，往往是非国家委托的个人行为。由于研究型专家只有文科知识背景，仅凭个人的力量，难以深入进行陶瓷显微及微观的研究。

总之，研究型的专家由于国家体制及工作性质的原因，到目前为止，他们在进入显微及微观世界寻找鉴定新方法的探索与研究中，不是职业的必须行为，而是个人的兴趣行为。

## 三、传统文化原因

中国是一个深受儒学文化影响的国家，与西方强调做事的文化不同，儒学文化的实质是更强调一种做人的文化，这种做人的文化导致了中华民族的现实性。在这种做人而不是做事的文化熏陶下，人们往往缺乏对自然科学的探索，缺乏对事物进行规律性总结的动力，缺乏将复杂事物与现象抽象概括出定义、定律的逻辑思考方式。

中国人最早使用计时工具，却始终停留在"日上竿头"、"掌灯时分"、"三更半夜"、"鸡叫两遍"等模糊计时的阶段，最终计时精确的钟表由西方人发明。

中国古人记录哈雷彗星数十次，但没有人总结它的运行规律，最终哈雷彗星由西方人命名。

中国的四大发明推动了世界文明的进程，却没有在明清时期继续发展成为近代科技。而西方的文艺复兴，产生了近代自然科学；启蒙运动推动了思想解放，推动了对知识的追求与探索，并发展了资本主义。

100多年前，中国就有人用放大镜观察陶瓷，但至今无人总结其纷繁复杂的陶瓷迹型特征与规律，并使之上升为理论。这些现象能说与儒学思想影响、缺乏形而上的逻辑思考和强调做人的文化无关吗？

## 四、传统思维原因

传统的思维模式，其主流是模糊性，重自省体验，强调直觉顿悟，这与科学要求的精确与量化的思维模式不相符。

中国传统的语言表达方式，文约义丰，反映了思维的模糊性、随意性、暗示性等。如老子曰："道可道，非常道。"庄子曰："言有尽而意无穷。"禅宗曰："不立文字，顿悟成佛。"

纵观中国古代思想著作，思想家提出的观点大都是个人对社会主观论断性的观点，属于个人悟出来的心得，而不是利用逻辑思维客观地揭示事物本质的规律。所以大多数古代学者给我们留下的知识和思想，都属于对社会现实的感悟与总结，缺乏对自然界定量的思考和客观的逻辑性推演。

中国传统的学习方式是死记硬背。古时的中国人自幼学习儒家经典，要求倒背如流，而理解方式也充满主观相对性，书中涵义只能自己意会，所谓"读书百遍，其义自见"，缺乏逻辑思维指导，缺乏量化思考，喜欢模棱两可。

而科学要求思维是要追求精确、量化的表达。

古希腊哲人苏格拉底提出热爱智慧、追求智慧的观点，亚里士多德最早提出逻辑的理论，毕达哥拉斯提出关于"数"的概念，德谟克利特提出关于"原子论"的概念。这些古代西方思想家的观点成为西方人求知求实精神与逻辑思维方式的思想根源。

此外，西方人的信仰和价值观的根基在于追求来自上帝绝对的、永恒性的真理，这就造成了西方人对唯一性真理的不断追求。

古希腊追求智慧以及基督教追求真理的精神使西方人习惯用逻辑思维认识和分析事物，要求对事物有客观确切的概念性认知，这为西方人在科学、文化、艺术等各方面的成就奠定了逻辑思维基础。

中西方思维方式不同，也直接表现为对事物的认知和处理方式不同：

中国的日晷、圭表反映时间阶段，西方的机械钟表精确到分秒；

中国的中医判断病情的性质，西方的医学锁定病人的病灶；

中国的饮食调味靠的是经验，西方的方便快餐靠的是标准。

正因为如此，在陶瓷鉴定领域，由于受传统思维影响以及缺乏科学普及的社会基础，中国人更喜欢没有具体标准，喜欢形象思维的眼学目鉴方式，崇尚个人的经验传授。

## 五、传统鉴定方法存在局限性

除了研究型专家，中国还有另一类专门从事鉴定的专家。他们主要用传统的眼学方法鉴定器物。这

separation is due to the fact that they are organized under a separate department from that of the authenticators.

Of course there are a small number of experts that are involved in research qualified with license to authenticate because they are free to transfer within the government departments and have acquired the license.

Amongst the research experts, the large majority belongs to University archeology faculties, others to provincial archeology research centers, with the remainder on staff at government-run museums. The focus of their work is research, not authentication, because the authenticity of what they are studying does not need to be questioned. Most of the pieces they encounter have already been authenticated, as they are predominantly from archeological excavation sites. Generally speaking, pieces from archeological discoveries can be directly associated with a specific time period. Therefore, further authentication is not required on such pieces. If they come across a site of unknown date, they can invite an associate expert from the authentication department to help establish the date, but there is no need to further determine its authenticity.

It is important to understand that these experts' entire earnings come from the government. Due to the code of ethics in their industry, it is impermissible to go outside of their work and engage in cultural relic trade. Therefore they lack the market experience of deciphering real from fake. They spend their energy on researching what is real, and thus have no need to recognize what is fake. As far as entering into the world of microscopy and microcosm on ceramics, it is not included within the scope of their job, and thus is not researched. Due to the nature of their work, an assessment of their ability to authenticate is not required by their respective departments or organizations.

As a result, these experts lack motivation to partake in research that is outside their respective filed. Their work does not facilitate pressure or urgency to further delve into microscopic study. If, however, they are found to be overly fond of authentication, it could be misconstrued as ignoring ones proper occupation and running irrelevant business on the side. Because of this, there currently is only very small portion of researchers who have become interested and have begun participating in relic authentication. Thus, as society starts to demand more research based authentication it will be coming not from government researchers, but from amongst the private sector. However, coming to a perplexing impasse whilst authenticating, a researcher relying on a liberal arts background and their own ability will meet difficulty in delving deeply into microscopic research of ceramics.

In short, due to the government systems in place and the nature of their work, research professionals have thus far been unable to establish microscopic study of ceramics as a viable method of research and authentication. Today, entrance into the microscopic and microcosmic realm to find new authentication methods and deepening research is not driven by the requirements of a profession, but instead has come from personal interest.

## 3. Traditional Culture Influence

Differing from the 'get stuff done' emphasized culture of the West, China is a country that has been deeply impacted by Confucian culture. Confucian culture emphasizes personhood, and thus is a very realist culture. Under the influence of this personhood vs. 'get stuff done' culture in the West, Chinese have often lacked an exploration of natural sciences, the motivation to summarize the laws of nature, and a methodology for logically explaining natural phenomenon.

Chinese were the first culture to track time, but always remained in a state of vague timekeeping. However, it was the West that invented the precise timekeeping of mechanical clocks.

Chinese recorded the passing of Halley's Comet more than forty times, but never summarized the laws of its movements. It was in the West in which the comet was named.

Through The Four Great Inventions (Paper, Printing, Gunpowder, and Compass) Chinese pushed forward the progression of World civilization, but failed to continue to develop these into new modern technology during the Ming and Qing Dynasty. The Renaissance in the West resulted in modern Natural Science and the Enlightenment pushed forward the liberation of thought and developed capitalism.

About one hundred years ago Chinese had already begun

些专家分属在省的文物鉴定站或国营文物商店之中。他们的日常工作主要是做鉴定，而不是做研究。

中国传统的眼学鉴定方法还没有被纳入中国高等教育体系，中国高等院校中至今还缺少有学位授予权的古陶瓷鉴定学科。因此，眼学鉴定方法主要靠师徒式、短训班式的模式传承。

20世纪80年代初，市场行家出身的国家级专家耿宝昌先生以短训班模式，向全国部分文物工作者讲授明清瓷器鉴定。

80年代末，耿老在讲课稿的基础上，整理出版了《明清瓷器鉴定》。这几乎是一本圣经式著作，因为至今无人敢说自己的书能出其右。《明清瓷器鉴定》从鉴定经验上来说，已经达到了难以企及的高度。

叶喆民教授在《明清瓷器鉴定》序言中说道："……过去有关古陶瓷学文献屈指可数，近代名著如《陶说》、《陶雅》、《陶录》、《说瓷》等虽然不失为一代名作，但在具体而微之观察方法尤感不足，今日幸得耿宝昌先生《明清瓷器鉴定》一书正式刊行，内容论述周详，剖析精微，足以启迪后学，发人深思，实属文物鉴定方面富有真知灼见、图文并茂之难得佳作。"[2]

叶教授评价过去的陶瓷鉴定书籍，认为其在具体而微之的观察方法上有不足，肯定了耿先生的"剖析精微"弥补了前人的不足。但耿先生的具体而微之的观察方法还是眼学的观察，观察所使用的放大镜以8—18倍的为主，仍是以器物的造型、纹饰、胎质、釉色、款式、工艺等要素作为鉴定断代的主要依据。

由于高等教育没有"鉴定"这样的专业及学科，80年代以前的鉴定人员，大多数都没有大学学历。就目前来说，有资格从事鉴定的专家，其构成比较复杂。有大学文凭的，大多是学历史或考古转行过来；没有文凭的，由于在文物部门工作，得到了组织上的重点培养，时间长了也可以评职称、当专家。即使像师傅带徒弟那样的短训班，国家也没有经常性地办下去，因为在八九十年代国家以经济工作为重心，对文化工作不够重视，文化层面的事国家顾不了太多，拖欠了鉴定专家许多学习账，以致有成就的鉴定专家都属于自学成才。

目前瓷器鉴定的方法，从专家们的论文和著作统计来看，主体上还停留在耿宝昌所代表的民国时代，还是以经验性鉴定的人证为主。另外，在眼学鉴定中，由于没有一个类似网球比赛中"鹰眼"机制的最终裁决，如果出现了专家之间对文物断代及真假判断不一致的情况，结果只能是谁说了都算，谁说了都不算。杭州南宋官窑博物馆的专家们关于该馆的镇馆之宝——长沙窑"壶王"之争就是典型的例子。

在一般情况下，我们无法评判专家们的水平高低，以及他们对器物鉴定的对与错，因为公说公有理，婆说婆有理。只有当一个造假者制造出的假货被专家断定为真品时，才能检验出专家的错误。

眼学鉴定专家，当他们双方的鉴定意见有冲突时，一方面缺少第三方充当"鹰眼"的仲裁者，另一方面，对他们之间相对立但又无法印证的观点，国家与社会选择了宽容与沉默。当然，这是无奈的选择。我们不能满足于眼学鉴定没有裁判的好处，而惰于创新。

## 六、入门要求原因

传统眼学鉴定方法的传授，不设学历门坎，即使小学学历的人也可以跟着师傅学鉴定。但学习显微及微观的研究与鉴定，门坎要求较高。因为显微及微观研究的目的，就是在传统眼学的基础上，将传统的定性研究变为定量研究。定量研究需要更多的高等教育中的理工科知识背景，因此进入量化研究的门坎高。

但是，只有走向定量的研究，才是真正走向科学的研究，定量研究实际上是与国际接轨的研究模式。要进行定量研究，走实证科学的道路，必然涉及数、理、化等理工科知识。因此，受过理工科高等教育的人进入显微及微观研究领域，要比没有受过理工科高等教育的人相对容易。

另外，眼学鉴定可以单枪匹马出成绩，而进入显微及微观领域研究，必须是团队合作才更容易出成绩。例如广州番禺博物馆和广州东方博物馆的联合研究团队，各类专业人士有：燃烧学博士、化工硕士，还有热处理、计算机、生物学、哲学、考古学、中文、经济学、社会学等本科及本科以上学历的专业人才。

## 七、小结

鉴于以上诸多原因，虽然显微镜的发明至今已有几百年的历史，但这项观察技术在陶瓷研究与鉴定领域的应用尚停留在起步阶段。

to use magnification to study porcelain traces, but until now no one has summarized the characteristics and laws of those traces or construct a logical theory. Can you say that these things have nothing to do with the influence of Confucianism, lack of logical thinking, and the emphasis of personhood culture?

## 4. Traditional Ideology Influence

Bemusement is mainstream in Traditional Chinese Ideology; relying in experiential introspection, emphasizing intuition and enlightenment. This does not suit the clear-cut and quantitative ideological model required in science.

Traditional language expresses: 'Words few, meaning abundant'; thus reflecting the bemused, capriciousness, and suggestible nature of traditional thought. These are all enforced in the teachings of Ancient Chinese Scholars such as Lao-zi, Zhuang-zi, Zen, and so forth.

A broad look into Chinese ancient ideological works shows that the majority of the viewpoints that Chinese thinkers held were personal subjective judgments on society's function, and were not utilizing logical thought to proclaim the essence of things. Therefore the majority of traditional ideologies passed down from ancient scholars are ponderings on societal issues, and lack reflective deductions, objective logical conclusions and firm realizations about the natural world.

Traditional Chinese learning is one of mechanical memorization, or learning without comprehension. In ancient times, starting at a very young age, Chinese students would memorize the Confucian Classics to a level of fluent recall, while the requirement for comprehension was simply subjective relativity. The innate meaning of books was left to be intrinsically grasped, as it was said, "Read a book a hundred times, meaning comes from within". The system was lacking in directed logical thinking, quantitative reflection, and was prone to equivocation. The requirement of science, on the other hand, is the pursuit of expressing thought both clearly and objectively.

Greek philosopher, Socrates, ardently pursued wisdom and established philosophy; Aristotle was the first to put forth the theory of logic; Pythagorean established his mathematic theorems; and Democritus developed the concept of atomic theory. It was on the foundations laid by these thinkers that the West gained its fervor for the pursuit of truth and wisdom.

In addition, Western religion and value system is based upon God's absolute and eternal truths, which in turn creates an unending drive to understand and uncover unique truths.

Ancient Greece's pursuit of wisdom and Christianity's pursuit of absolute truth created a habit of using logical thinking to recognize and analyze the world around. It was on the cornerstone of logical thinking and a need for absolute and objective conceptualization that the West established science, culture, arts, and other facets of its society.

Chinese and Western differences in ideology can directly be seen in their understanding and handling of the world around them:

Where Chinese used the approximation of a sundial to tell time, Westerners measure each second with a mechanical clock.

Where Chinese Medicine determines the relativity of a sickness, Western Medicine locks in and focuses on the disease.

Where Chinese cooking draws from experience for flavoring, Western fast food relies on standard measures for convenience.

This is precisely why Chinese have remained idle and strayed away from the precision of microscopic study, choosing instead to rely on approximate intuition for authentication.

## 5. Traditional Authentication Method Has Limitations

Aside from the research professional, China has a division of experts that are dedicated solely to authentication. They primarily use the traditional intuition approach when authenticating articles. These experts are spread throughout provincial level relic authentication stations as well as government run relic brokerage houses. The primary focus of their job is to provide authentication and does not include research.

Chinese traditional intuition authentication method is not

a curriculum based scholastic field, and to this day there are very limited amount of university level courses offered in the field of antique authentication. Because of this, the intuition authentication method is passed on through an Apprentice-Master model or short-term course model.

In the Early Eighties an expert, Mr. Geng Baochang, who was formerly a market connoisseur, taught a short-term course to some of the cultural relic workers on Authenticating Ming and Qing Dynasty Porcelain.

In the Late Eighties, based on his lecture notes, Mr. Geng published the book *Ming Qing Ci Qi Jian Ding* (Authentication of Ming and Qing Porcelain). It has become the 'Bible' of authenticating, as to this day there is still no one who has been so moved as to say their book surpasses that of Geng Baochang. Speaking merely of authentication experience, *Ming Qing Ci Qi Jian Ding* is at a level extremely difficult to reach.

Professor Ye Zhemin, in the forward to *Ming Qing Ci Qi Jian Ding* stated, "Publications on antique porcelain can easily be numbered: Even though *Tao Shuo*, *Tao Ya*, *Tao Lu*, *Shuo Ci* can be considered as masterpieces, I feel they are lacking in minute observation methods. We are fortunate to have Geng Baochang's *Ming Qing Ci Qi Jian Ding*; a very formal publication, with comprehensive contents, intricate detail, and a substantial push for further study and deeper understanding; a truthful piece on relic authentication; richly profound, with exquisite illustrations and text; a true masterpiece." [2]

Prof. Ye's pointing out the lack of 'minute observation method' in other records and appreciation of Mr. Geng's 'intricate detail' shows his belief that Geng's method makes up for the inadequacies of those that came before. However, Mr. Geng's 'intricate detail' observation is still intuition based, utilizing 8-18X magnifications for observation. Therefore, authentication was primarily founded on the piece's shape, pattern, body composition, glaze color, marking, and production technique.

Because 'authentication' is not contained within the Higher Education system, authentication experts before the 80s are without corresponding academic credentials. Currently, the system for qualifying experts to take part in authentication is quite complicated. For those who have university diplomas, majority are history or archeology majors that have changed professions. Those that don't have university diplomas have stayed long enough within the relic department, receiving training internally, to take part in appraisal and be called 'expert'. However, the government has drifted away from the traditional Apprentice-Master model of the past. During the Eighties and Nineties the government placed a larger importance on jobs within the financial sector than those within the culture sector, so the jobs within the cultural areas began to fall by the wayside. The study resources for authentication experts slowly dried up, leading to the majority of their studies being self-conducted.

Currently the experts' method of authenticating of porcelain, according to articles and publications they have written, has remained in the Republic of China Era intuition based human witness system represented by Geng Baochang. Additionally, this naked-eye intuition based authentication is without an authoritative standard 'Hawkeye' system similar to that made popular in tennis and other sports utilizing instant replay. If there is a discrepancy between experts as to what is real and what is fake, there is no underlying standard to say what the ultimate truth is. A prime example of this is the Changsha Kiln 'King of Kettles' dispute from the Southern Song Dynasty Guan Kiln Museum of Hangzhou.

For the most part, there is no way of measuring the experts' level of expertise. There is no way to judge their authentication as every one has their own standards for authentication, and every one can hold their own opinion as to what is authentic. It is only when a forger brings forth a known forgery and the expert authenticates it as a real piece that we would know the mistake of said expert.

When these naked-eye intuition based experts come to an impasse on a specific piece, there is no third party 'Hawkeye' to arbitrate the decision. Additionally, since there is no way to corroborate either side's viewpoint, the country and society chooses to accept the difference and remain silent. Of course, this is a choice forced by lack of alternative. In this case, experts are content with the intuition-based method without a final judge and lack motivation to push for innovation.

## 6. Threshold Influence

The impartation of the intuition-based method of authentication is free of an academic based threshold; even a pupil with only primary school education can enter into an apprenticeship program. However, the threshold to cross in order to carry out authentication research through microscopic and microcosmic observation is relatively high. This is because microscopic and microcosmic research builds over the top of the foundation of the intuition-based method, which is to move from studying the nature of an item to studying the quantifiable properties of that item. Quantitative research is rooted in Higher Education as well as scientific and technological knowledge, which is precisely why the threshold to enter this field of study is considerably higher.

In as much as moving towards quantitative research is moving towards a more scientific style of research, it is also the integration into the internationally acceptable model of research. To take the route of scientific research, it is imperative to travel along the road of science and engineering fields of Mathematics, Physics, Chemistry, and so forth. It is only when microscopic and microcosmic research is founded upon these technological pillars that crossing the threshold is relatively easy.

Additionally, the intuition-based model of authentication can be carried out single handedly, whereas a diverse research group or team most efficiently and easily carries out microscopic and microcosmic research. For example, the research team at Guangzhou Panyu Museum and Guangzhou Oriental Museum is one of varying and diverse educational backgrounds including: Doctorate of Combustion Studies, Master of Chemical Engineering, Bachelors of Thermal Mechanics, Bachelors of Computer Sciences, Bachelors of Biology, Bachelors of Philosophy, Bachelors of Archeology, Bachelors of Chinese Language, Bachelors of Business, Bachelors of Sociology, and so forth Bachelor level or higher educational backgrounds.

## 7. Conclusion

In light of the above stated barriers, even though the development of microscopes and microscopic research has enjoyed several hundred years of history, the application of this form of observation within the realm of research and authentication of ceramics has until now remained in the preliminary stages.

# 第四节
# 陶瓷鉴定方法的发展演变

陶瓷鉴定的最终结果，是要拿出证据证明被鉴定器物的真假，而不是简单的以理服人。在陶瓷鉴定的证据链中，物证要比人证更加具有说服力，因为物证更接近客观事实。所以说，陶瓷鉴定方法的发展与演变过程，本质上就是陶瓷鉴定证据链不断完善的过程，就是从人证走向物证、旁证走向自证的过程。

那么在陶瓷迹型学诞生之前，陶瓷鉴定的证据是怎样的呢？

## 一、口说为凭经验证据

20世纪50年代以前，古陶瓷鉴定主要使用传统的眼学方法：通过眼观、手拎、耳闻，对陶瓷的胎质、釉色、纹饰、口足、款识等方面综合观察，凭经验和凭一些未被明确发掘时间、发掘地点及年代依据的传世标准器，通过比对来判断被鉴定器物的真伪、年代及窑口。传统鉴定的技能，来自鉴定者平时的多看、多比、多问所积累的鉴定经验，来自师傅对徒弟的传授，来自对书本知识的学习。

民国以前，用标准器进行比对鉴定会有一定的鉴定困难，因为所运用的标准器系统存在以下问题。首先，由于缺乏明确发掘时间、发掘地点及年代依据的记录，一些传世的标准器本身就存在一定的疑问。即使由科学考古发掘出来的来自窑址、窖藏或沉船的器物，要获得其准确的年代信息也不容易。而且，标准器毕竟数量有限，无法概括所有该类型古陶瓷的特征。当用这些标准器去给非对应的器物进行比对鉴定时，很有可能会产生判断的偏差。所以常有鉴定专家给出"没见过"的评论，标准器的短缺给传统的眼学鉴定方法带来了难以突破的障碍。所以，与标准器比对的鉴定方法，得出的结论是或然的。

师傅带徒弟的传授模式，在封建社会中，由于有着"教会徒弟，饿死师傅"的传统观念，师傅是不轻易把绝招交给徒弟的。因此，许多师傅的毕生经验就有可能失传。再者，那时候也没有评判鉴定师傅对错的机构。所以，正确的经验和有问题的看法会一同传承。

几千年来，中国有关陶瓷鉴定的书籍极少，这与中国封建社会不重视手工业阶层有关。比如，元朝将人分成十等，七匠八娼，工匠仅比娼妓高一等。正因为这种歧视工匠的传统观念，尽管中国制造陶瓷的历史有几千年，但论瓷之书与制瓷的历史相比，悬殊太大。正如《饮流斋说瓷》中的记载："吾华诸美术以论书画之书为最多，以其与文人气相近也。若刻印，若范铜，则稍罕矣。而论瓷之书尤寥寥若晨星。"[3]直到清代，这种不被重视的情况也未得到改观。

再者，古人书籍论陶瓷鉴定，更多的是描述与比喻，容易产生多义。比如明代曹昭著的《格古要论》一书说道："出北地，世传柴世宗时烧者，故谓之'柴窑'。天青色，滋润细媚，有细纹。多足粗黄土。近世少见；"[4]民国时期的《饮流斋说瓷》论道："古瓷尚青，凡绿也，蓝也，皆以青括之[5]。"然而，前人文中提及的天青色、绿色、蓝色究竟是什么颜色，每个人的理解都会不同。

尽管传统的鉴定方法有自身的优势，但这种鉴定知识是经验性、封闭性、个人化、孤立化和碎片化的观点，属于主观的经验性证据。

## 二、客观实在物证证据

### （一）二重证据法

从20世纪50年代开始到80年代，随着我国陶瓷考古的发展，古陶瓷鉴定开始应用"二重证据法"，陶瓷鉴定的证据得到了进一步的完善。

20世纪20年代，学者王国维提出，要考证古代历史文化，需要综合利用古文献记载和从考古遗址发现的实物，也就是"纸上之材料"与"地下发现之新材料"二者互相

# Section 4:
# Evolution of Ceramic Authentication Methods

Ultimately, the authentication result of a piece under review must be given in terms of evidence, and not simply just a set of theorizations. In the chain of authentication evidence, material evidence is considered to be more compelling than that of human witness; this due to material evidence better representing objective fact. This is to say, the process of change and development to authentication methodology is essentially the unending progression along the evidence chain. It is the move from human witness to material evidence, then from circumstantial to self-supporting proof.

So, before the emergence of Ceramics Trace Model Study, what were the ceramic authentication evidences?

## 1. Oral Account Conveying Experiential Evidence

Until the 1950s, antique ceramic authentication primarily utilized a traditional intuition based method: Using vision, handling, and listening to comprehensively observe the body, color, pattern, shape, marking, and craftmanship. This observation would be compared with experiences of pieces of not necessarily known time and excavation place or solid provenance handed-down pieces to ascertain the authenticity, date, and kiln of the piece in question. The accuracy of this method relies on the amount of pieces seen and compared by the appraiser—essentially their personal experience, as well as coming from the passage of knowledge from master to pupil, and from the studying of published knowledge.

Up to the Republic Era (1911-1949), pieces used for comparison in authentication had some difficulties, because the sample piece system used had some inherent problems: First, due to the lack of known excavation time and place or date, some handed-down pieces themselves had questions. Additionally, discerning dating information on pieces that came from archeological excavations of kiln sites, hoards, and shipwrecks comes with difficulty. Also, 'sample pieces' were very limited, and there was no way to include the characteristics of all types of antique ceramics. When using these samples to conduct comparative appraisal, it is very common to come across inconsistencies in judgment. Therefore, it is common to have an appraisal expert simply say "Never seen anything like it" as a verdict. The small number of samples creates a burdening handicap in the traditional intuition based authentication method. Thus, the conclusions drawn from this sample comparison authentication method are similarly encumbered.

The Master-Pupil system for disseminating knowledge within a feudal society, due to the "If the pupil masters, the master will starve" traditional mentality, it was no easy matter for a master to pass on the tricks of the trade. Because of this, the full extent of the master's lifetime of experience remained stifled. Moreover, at that time there was no mechanism for passing judgment on the master if their authentication was out of line. Therefore, correct experience as well as problematic views were passed down at the same time.

For thousands of years, very limited publication had been written in regards to ceramic authentication. This is due large in part to Chinese feudalist society not placing a high value on the appreciation of handcrafts. For example, the Yuan court spilt society into ten strata. Seven being artisans and eight being prostitutes, craftsmen were seen as only one level higher than streetwalkers. It is specifically due to this kind of traditional concept discrimination that even after thousands of years of ceramic crafting in China, there is a great disparity in the amount of books discussing ceramics and the craft. Just as it says in *Yin Liu Zhai Shuo Ci* (Yin Liu Zhai Speaks on Ceramics), "Most of our splendid works are

释证，称"二重证据法"。同期，陈万里先生将王国维的观点付诸实践，考察龙泉窑，带回了一系列有出土时间、出土地点的陶瓷标本。之后，陈万里先生在其论文中指正故宫关于瓷器照片实物定名不准确的问题，指出古陶瓷鉴定只是参考古文献，而不参照考古出土标本，这样判断窑口是不准确的[6]。

陈万里先生是我国近代第一个在古陶瓷研究鉴定领域中，引入现代考古学的方法，并将其付诸实践的人。陈万里先生的贡献，是证实了用来作比对鉴定的标准器有明确的发掘地点与年代依据，这就改进了陶瓷鉴定的证据，改进了第一阶段的传统鉴定方法。用这样的标准器作比对鉴定，结论就具备了实然性。

新中国成立以后，随着考古的发展，窑址不断被发现，大量可供鉴定比对的标准器面世。从此，"二重证据法"正式被引入到古陶瓷鉴定中，一直沿用至今。但"二重证据法"在当今陶瓷鉴定实践中遭遇了严峻的挑战。仅用眼学和文化历史知识去比对被鉴定器物与标准器之间的共同性，这种鉴定方法的准确率已经大为下降，其主要原因来自于现代仿品的冲击。现代仿品不断缩小与标准器的差距，使经验性的眼学鉴定越来越吃力。

## （二）科技鉴定

随着当今陶瓷收藏的兴起，赝品泛滥，人们逐渐把陶瓷鉴定的希望转向了科技鉴定。

20世纪20年代，我国冶金学家和陶瓷学家周仁先生开创了中国古陶瓷科技鉴定的研究。之后中科院上海硅酸盐研究所、中科院等机构将古陶瓷科学技术研究不断深化、系统化[7]。20世纪80年代至今，科学技术被更加广泛地应用到古陶瓷鉴定中。古陶瓷科技鉴定方法主要是分析陶瓷表面、断面的显微结构，分析陶瓷各项理化性能和年代的测定研究等[8]。

古陶瓷科技鉴定中常用的科学技术包括：化学分析法和仪器分析法。

### 1. 化学分析法

化学分析法是以陶瓷的化学反应为基础来定性或定量研究其化学组成的分析方法。

化学分析方法对设备要求简单，结果准确，适用范围较广。但化学分析难以对陶瓷中微量和痕量元素组成的定性和定量做出准确分析，只适用于分析陶瓷中的主要元素，而且对样品有一定破坏性[9]。

### 2. 仪器分析法

仪器分析法主要包括：热释光测年法、元素分析法和其他成分分析法。

#### （1）热释光测年法

热释光测年法是利用晶体在受自然界辐射作用后积蓄的能量在加热过程中释放出一种磷光的物理现象来进行断代的技术。

热释光测年法的优点在于证据来自器物本身，不需要任何比对参照便可测定现代陶瓷和古代陶瓷，但缺点是对样品有一定的破坏性[10]。

国际上普遍认同热释光测年法。

#### （2）元素分析法

元素分析法主要利用现代仪器分析陶瓷的胎、釉、彩中所含的常量元素、微量元素和痕量元素的种类、含量及各元素之间的比例关系等[11]。

元素分析法在模式上采用从已知探未知的方法，根据相应时代、地域古陶瓷样品的元素组成特征和变化规律对未知样品进行分析判别。根据统计学原理，要让这种方法的判别依据具备科学性和准确性，要求必须掌握足够数量的有明确发掘地点、可靠地层和年代依据的古陶瓷标本的元素组成信息，建立数据库[12]。

元素分析法主要包括仪器中子活化分析、X射线荧光光谱分析、电子探针能谱分析等方法。

X射线荧光光谱分析是通过X射线的激发，对被测元素原子的内层电子发出X射线光谱线的波长和强度进行定性和定量分析的方法。X射线荧光光谱分析可分为能量色散X射线荧光光谱分析和波长色散X射线荧光光谱分析。其中能量色散X射线荧光光谱分析在古陶瓷科技鉴定领域中的应用更为广泛和深入，也是目前唯一被成功应用到古陶瓷无损鉴定的一种成分分析技术[13]。

这种方法灵敏度较好，分析速度快，自动化程度高，不会对样品造成破坏，但对原子序数较低的轻元素灵敏度低，不能区分元素的化学种类，而且检测时易受干扰[14]。

仅仅依据元素分析法所得出的数据来判断陶瓷真假还是不够的。现实中有位高仿专家甚至说，由他配方生产出的高仿品已能过机器检测关。

discussing paintings and other scholarly pursuits. There are a handful about stamp carving and bronze sculpting, but works discussing ceramics are few and far between." [3] All the way through the Qing Dynasty, this sort of discrimination did not undergo any drastic change.

Furthermore, the ancient scholarly works that do exist are merely a series of descriptions and analogies, easily giving rise to ambiguity. As seen in the muddled description in the famous Ming Dynasty record *Ge Gu Yao Lun*: "Fired in the north, during the reign of Cai Shi Zong Emperor, thus called 'Cai Kiln'; sky-blue, refined and charming, with fine crazing; multi-footed with coarse yellow clay; and rarely seen recently." [4] We also see confusion discussed in the Republic Era publication: *Yin Liu Zhai Shuo Ci*, "On ancient porcelain 'qing' color can be anything greenish or bluish [5]." The historically confusing use of the word 'qing' to describe green, blue, and so on; with every person's understanding of what is actually 'qing'; is just one example of the perplexities seen in traditional thought.

Even though the traditional authentication method has its inherent merits; its authentication knowledge base is experiential, inaccessible, individual-centered and fragmented in its viewpoints; reflecting its observer experience based witness as evidence.

## 2. Objective Material Evidence

### A. Double-Evidence Method

From the 1950s to the 80s, in accordance with the developments in Chinese archeology, antique ceramics appraisal began to utilize the 'Double-Evidence Method', bringing ceramics appraisal evidence one step further along the evidence chain.

In the 1920s, scholar Wang Guowei brought forth the idea that when distinguishing ancient cultures, there is a need for a comprehensive utilization of both ancient text and records in addition to archeological excavations. This is to say that the combination of 'information on paper' and 'information below the ground' can be used as mutually affirming evidence called 'Double-Evidence Method'. At this time, Mr. Chen Wanli put Mr. Wang's viewpoints into practice

on an examination of the Longquan Kiln; bringing back with him an entire series of known time and location excavated samples. Afterwards, Mr. Chen in his dissertation of the findings, pointed out inconsistencies in the Palace Museum calling on pictures of certain pieces; saying that ceramic authentication by only considering ancient writings and not giving thought to archeological samples for determining a piece's kiln origin is not precise [6].

Mr. Chen was the first to enter into the realm of antique ceramic authentication research in China, ushering in a new archeology research method, and putting it into practice. He was dedicated to using clearly defined samples of known excavation record for carrying out comparative authentication; thus transforming authentication evidence and beginning a new period in traditional authentication methodology. When using these types of samples for comparative authentication, conclusions can possess a fundamentally concrete key element.

After the establishment of the P.R. China, along with the development of the archeology field, increasing numbers of kiln sites were continually discovered, and new 'sample pieces' began to surface. From then on, the 'Double-Evidence Method' officially entered into the antique authentication world, and has been utilized until today. However, this Double-Evidence Method, within the present landscape of ceramic authentication, has run across a gauntlet of difficulties. Using only intuition and cultural understanding, the percentage of accuracy in comparing the similarities of the piece in question against a sample as an authentication method has seen a sharp decrease. This is owed to the continuous pestering by the modern forgery industry. The discrepancy between modern fakes and their authentic counter parts has become increasingly minute, thus eating away at the effectiveness of experiential intuition based authentication.

### B. Technological Authentication

In conjunction with the rising interest in ceramics collection and flood of art forgeries, people have begun to shift their hopes of ceramic authentication towards technology.

In the 1920s, Chinese metallurgist and ceramicist Zhou Ren set out on Chinese antique ceramics technological

由于这种方法需要比对古陶瓷标本的元素组成数据，而目前古陶瓷数据库还不够全面和完整，所以使用元素分析法得出的结论在鉴定古陶瓷时也只能作参考。

### 3. 小结

以上谈到的科技鉴定古陶瓷方法都有它们共同的局限性：

（1）由于仪器的检测精确度有一定误差范围，无法精确到某个皇帝当朝的年代。

（2）对于元素分析法，其所建立的数据库的全面性和完整性也会影响判断的准确性。

（3）鉴定古陶瓷的仪器操作专业要求高，对广大收藏爱好者而言缺乏易懂的直观性，且购买费用昂贵，难以普及。

## 三、物证系统——陶瓷迹型学创立

那么，有没有更好的方法，既能让陶瓷鉴定无损文物，又能让收藏大众直接参与呢？可以说，陶瓷迹型学的诞生，既开创了问题的解决之道，又使科研成果民生化、市场化。

### （一）陶瓷迹型学创新了观察角度

如果说，传统鉴定陶瓷是第一只眼睛断真假，科技鉴定陶瓷是第二只眼睛断真假，那么陶瓷迹型学就是第三只眼睛断真假。

新的观察角度，就是全新的认识陶瓷真伪的角度，它延伸了我们的观察器官，在放大500—600倍的显微层面展开，开拓了陶瓷鉴定的新领域，为陶瓷鉴定拓宽了视野，提供了新方法。

### （二）陶瓷迹型学创新了观察方法

马士宁三维成像法，首次将鲜为人知的陶瓷釉表面的三维形貌展现出来，为区别自然风化的量变特征与人为做旧的质变特征，提供了新的技术手段，从而增强了鉴定证据的说服力。

### （三）陶瓷迹型学创新了客观物证系统

陶瓷迹型学观察了大量具有明确发掘时间、发掘地点和年代依据或有明确传世记载的标本，将其迹型进行分析研究，归纳、提炼、建构迹型标准图谱模型。标准图谱模型是客观的物证系统，提供了更集中、更全面、系统化的证据，将之用于比对鉴定，类似于人类利用DNA基因图谱的亲子鉴定及司法鉴定中的笔迹鉴定，这种直观比对鉴定，系统化、形象化，易操作、易普及。

### （四）陶瓷迹型学创新了鉴定依据

陶瓷鉴定准确率的提高，有赖于证据链的不断完善。民国时期鉴定水平高于之前的鉴定水平，因为有明确了发掘时间、发掘地点和年代依据的标本，这些具有客观实在性的标本强化了鉴定依据，使鉴定结果更准确、更具鉴定说服力。但是，在如今高仿制瓷的冲击下，标准器与伪品之间的差距在缩小，眼学目鉴难以分辨。

于是，陶瓷迹型学改变了鉴定依据，从标型鉴定依据改成了迹型鉴定证据，从标准器比对改成了迹型标准图谱模型比对，将传统鉴定储存在头脑中的表象记忆证据移植到现实中来，这不仅使真品与伪品更容易分辨与判别，而且增强了客观证据的说服力。

陶瓷鉴定的迹型证据还将科技鉴定的抽象化证据转换为形象化证据，可让收藏爱好者直接参与鉴定，直接验证物证的可靠性。

以陶瓷迹型及它们之间关系的规律作为证据，可以从有限推及到无限，可以弥补眼学鉴定中标准器不足的遗憾。

### （五）陶瓷迹型学创新了鉴定研究模式

传统的陶瓷鉴定知识的学习与研究，是以一个人研究为基础的，因而常常有所谓不轻易示人的独门秘籍的种种传说。而陶瓷迹型学的研究是建立在文、理工科等多学科团队研究基础之上的。这是科学量化研究的基本组织模式。发挥多学科共同研究的综合优势，是未来陶瓷迹型研究的发展方向，也是社会与科学发展的必然要求。

### （六）陶瓷迹型学创新了鉴定表达模式

陶瓷迹型学鉴定将以迹型标准图谱模型比对，以书面的逻辑论证、量化的数据，以科学的鉴定报告形式，阐述鉴定结果。

authentication research. Subsequently, the Chinese Sciences Academy's Shanghai Silicates Research Institute and its other associated organizations began continually deepening and systemizing research into the science and technology of ancient ceramics [7]. From the 1980s until today, the use of science and technology has become increasingly widespread in the field of ceramics authentication. Antique Ceramics Technological Authentication methods primarily consist of: glaze surface analysis, microscopic structure survey, and determining the physical and chemical properties and age of the respective phases through analytical research, to name a few [8].

Antique Ceramics Technological Authentication consists of the following two fields: Chemical Analysis Method and Instrument Analysis Method.

### I. Chemical Analysis Method

Chemical Analysis Method is founded on researching the type and amount of chemical reactions on ceramics to analyze their chemical composition. Chemical analysis has relatively simple technical requirements, offers accurate results, and is suitable for a wide range of materials. The difficulty arises from its inability to accurately analyze the amounts and nature of small and trace element compositions, and is only suitable for analysis of the primary components. Additionally, it leaves a certain amount of damage to the test piece [9].

### II. Instrument Analysis Method

Instrument Analysis Methods primarily consist of: Thermoluminescence Dating (TL) and Elemental Analysis including other composition analysis methods.

#### a. TL Dating Method

TL Dating is the determination, by means of measuring the accumulated radiation dose during the time elapsed since a material containing crystalline minerals was last fired.

The benefits of TL testing is that it offers evidence from the actual test item and does not require any sort of comparison to determine the age of modern or ancient ceramics. The drawbacks however, are that it also damages the piece in question [10].

Internationally, TL is the most accepted means of dating.

#### b. Elemental Analysis Method

Elemental Analysis Methods all essentially use modern scientific instruments to analyze the types, amounts, and relative ratios of major, minor, and trace elements within the porcelain body, glaze, and coloration [11].

Elemental Analysis Methods utilize a database of known specimens of ancient ceramic elemental compositions to compare against test pieces in order to analyze and determine relative age. In accordance with statistical analysis, in order to have scientific and accurate results, this method must utilize a substantial amount of elemental composition information from sample ceramic specimens of known excavation site, reliable strata, and identified date for construction of its database [12].

Currently, the most predominantly used Elemental Analysis Methods are: Instrument Neutron Activation Analysis, X-Ray Florescence Method (XRF), and Electron Probe Power Spectrum Analysis.

The XRF method utilizes X-rays or gamma rays to stimulate the atomic particles within a certain material and then reading the emission of the characteristic florescent X-rays from that material. There are two main types of XRF methods: Energy-dispersive X-ray spectroscopy and Wavelength dispersive X-ray spectroscopy; with Energy-dispersive being more widely used in the realm of antique ceramic authentication. It is currently the only widely used non-abrasive Elemental Analysis Method [13].

The sensitivity of this method is relatively high, the speed of analysis is relatively fast, the degree of automation is high, and it has no known harm to the piece being analyzed. However, for elements with low atomic numbers, the sensitivity is quite low, with an inability to distinguish the elements' chemical classifications. Additionally, the testing is very easily disrupted [14].

Unfortunately, the elemental composition data discovered by the Elemental Analysis Methods is still not yet sufficient. Additionally, there is already a forgery expert who claims that the pieces he has crafted can already pass instrument testing.

Due to the large amounts of antique ceramic sample data

### （七）陶瓷迹型学创新了传播模式

传统的陶瓷眼学方法，以师傅带徒弟的传授模式为主。在旧时"教会了徒弟，饿死了师傅"的传统观念影响下，不能保证所有的师傅都会将看家本领及绝招统统传授给徒弟，有时候因此会造成一代不如一代的现象。而且师傅只能传授经验，不能传授阅历，师傅的成功主要源于个人的阅历修炼与悟性。

陶瓷迹型学以概论式的阐述，建立了陶瓷显微迹型研究的理论框架，形成了系统的理论学说。该专著可被纳入学制式高等教育体系进行统一传播，这改变了千百年来陶瓷鉴定知识个人非统一的传播模式，这是陶瓷鉴定知识的传播模式中意义深远的一场革命。

### （八）陶瓷迹型学创新了显微鉴定的理论体系

陶瓷迹型学是建立在显微研究基础之上的，陶瓷的显微研究在10年前就有人涉猎，但至今仍处于"散论"阶段，未达到"概论"阶段。因为概论需要量化研究，需要科学方法，需要定义概念，需要从现象到本质分析概念以及分析概念之间的关系，从而总结出规律，建立理论框架，形成独树一帜的系统学说。

陶瓷迹型学对陶瓷显微研究与鉴定作了概论式的系统阐述，首次定义了陶瓷迹型的类别、种类、单项、风化强度、风化作用强度、风化绝对时间、风化相对时间等重要概念，并归纳总结了迹型之间关系的规律，强调让事实说话，强调逻辑推理，强调实验证明，创造了新的鉴定理论体系。

## 四、小结

陶瓷鉴定方法的发展演变，就是陶瓷鉴定证据链不断完善的过程。从主观的经验性证据发展到客观的物证证据，再从客观的物证证据发展到陶瓷迹型标准图谱模型。标准图谱模型是客观、广泛、典型、系列的物证系统。这一发展过程就是中国陶瓷鉴定的发展史。《陶瓷迹型学概论》为中国陶瓷鉴定发展史增添了新的历史篇章。

清代以前是中国陶瓷鉴定的第一个阶段，民国至今是以陈万里为代表的"二重证据法"与科技鉴定并存的时代，这是陶瓷鉴定发展的第二个阶段。陶瓷迹型学的诞生，是第三只眼睛看陶瓷的真伪，它为中国陶瓷的鉴定增添了新生力量。

needed to carry out the comparative analysis, and as there is currently no complete or substantial database, using Elemental Analysis to carry out conclusive authentication of antique ceramics can only be used as supplementary at this current time.

### III. Summary

The above-mentioned technological methods for authenticating antique ceramics share some common disadvantages:

a. Due to the degree of error present in instrument testing, there is yet an exact means of precisely determining in which Imperial Reign within a certain Dynasty a piece was made.

b. In reference to Elemental Analysis Methods; the totality and completeness of the constructed database directly affects the accurateness of date determination.

c. Technological authentication methods lack easy-understandability and intuition: That is to say that the degree of professional requirement for instrument management is high, and apparatus purchase very expensive. Currently, the average collector is not able to participate in instrument authentication, and it lacks a certain level of user-friendliness.

## 3. Material Evidence System – Established by Ceramics Trace Model Study

Now, is there a better method available; one in which authentication is not only analytical, but also able to be directly participated in by the average antique collector? We can say that the emergence of Ceramics Trace Model Study is the pathway to resolving this problem; bringing the fruits of scientific study to the layperson and the market.

### A. New Observation Perspective

If we are to say that traditional authentication is the first perspective in authentication, technology offering the second, then Ceramics Trace Model Study can be considered as the third judge in the determining of real from fake.

Using a new angle of observation offers us a completely new viewpoint when going about authentication. This approach calls upon observation tools to expand our inspection of ceramics to a microscopic magnification of 500-600 times, opening a new realm of ceramic authentication by bringing forth an entirely new approach.

### B. New Observation Method

The Bunney Three-Dimensional Imaging Method marks the first manner in which we can begin to understand the three-dimensional appearances of ceramic glaze. It helps to delineate the stark contrast between the quantitative changes of natural weathering and the sudden changes brought on by human forced aging. A new technique, it fruitfully adds plausible evidence to the authentication process.

### C. New Material Evidence System

Through the observation of numerous samples with known excavation record and age or concrete provenance, and by using Ceramics Trace Model Study we have analyzed, researched, classified, and refined ceramic traces to construct a Standard Trace Atlas Model. Standardized atlas models are objective material evidence systems offering more concise, holistic, and systemized evidence. This is then used in comparative authentication, similar to that of Human DNA paternity testing or handwriting analysis used in courtrooms. This sort of comparative authentication is systemized, comprehensive, user-friendly, and easily standardized.

### D. New Authentication Foundation

Raising authentication accuracy relies directly on the continual improvement of the evidence chain. The level of authentication during the Republic Era was higher than that of periods before simply due to the fact that samples of known excavation record and age were used. This new injection of objectively factual samples strengthened authentication evidence; bringing more accurate conclusions and more persuasive appraisals. However, under the surging stream of new forgeries, the line between real and fake has become increasingly fine, and intuition based appraisers are struggling to differentiate.

Accordingly, Ceramics Trace Model Study has upgraded authentication evidence from sample model to Trace Model authentication proof, from sample piece comparison to Standard Trace Atlas Model comparison. Thus, it transitions us from the traditional intuition and memory based evidence into a realm of concrete reality. This not only assists in the differentiation of real and fake, but increases the credibility of the objective material evidence.

Ceramics Trace Model Study evidence is also able to restructure the abstracted verification of technological authentication into a more comprehensive set of evidence; allowing hobby collectors to directly participate in and personally test the reliability of the proofs.

Utilizing the laws that govern the ceramic traces and their interrelationships, we are able to move from limits of the inadequate samples used in intuition based appraisal into an unrestricted arena of authentication.

## E. New Authentication Research Model

Traditional authentication is founded upon solitary studies and research, and is often shrouded in mysterious axioms and adages propagated from one generation to the next. Ceramics Trace Model Study, on the other hand, is constructed on the foundation of a interdisciplinary research team consisting of areas of study ranging from language to physics and chemistry to archeology, just to name a few. This is the essential groundwork of a scientific quantitative research model. The benefit is putting forth comprehensive conclusions of multi-dimensional scientific research; which is precisely the future course of Ceramics Trace Model Study, as well as the basic demand of society and science.

## F. New Authentication Expression Model

Ceramics Trace Model Study, through the Standard Trace Atlas Model comparison, will put forth logical and quantitative evidence based authentication outcomes that are akin to scientific reports.

## G. New Authentication Propagation Model

Traditional intuition based appraisal methods are primarily passed on through the master-pupil model. Under the influence of the old adage "Once the student masters, the master will starve", there is no way to ensure that the full experience and repertory of the master will faithfully be passed down to the apprentice; lending to the phenomenon of each generation being lesser than the prior. Additionally, masters can only ever speak of their own experiences, without being able to fully pass on their own intuition. Thus, an expert's success relies wholly on his or her own experience and comprehension.

The easy-to-grasp nature of Ceramics Trace Model Study has erected the logical framework of microscopic trace research, creating a systemized medium for furthered study. This allows for the passage of knowledge to be carried out by means of Higher Education curriculum. This completely revolutionizes the thousand-year-old custom of passing ceramic appraisal knowledge via oral propagation.

## H. New Microscopic Authentication Theory System

Ceramics Trace Model Study is founded in microscopic research, which had already been used in ceramics study over a decade ago. However, until now, this research remained in a period of conceptualization, without concrete theories. This is because theorization requires quantitative research, scientific methods, defined concepts, and needs to move from phenomenon to essential characteristic analysis and analysis of the relationships between these concepts. From which, patterns and laws must be extracted, a logical framework constructed, and systemized experiments carried out.

Ceramics Trace Model Study has created a system for conveying the concepts of ceramic microscopic research and authentication. This is the first time that ceramic trace categories, trace types, and individual traces as well as weathering intensity, weathering severity, absolute weathering time, and variable weathering time have been defined and categorized. Additionally, the laws that govern the relationships between these traces have been surmised. This method stresses that facts speak for themselves, it features logical reasoning and also emphasizes proof through experimentation; thus creating a new body of authentication theories.

## 4. Summary

The evolution of ceramic authentication methodology is simply the process of continual improvement to the evidence chain. It is the progression from observer experience based witness, to objective material evidence, and from material evidence to the Ceramic Standard Trace Atlas Model. An objective, encompassing, definitive, and categorical material evidence system characterizes each Standard Trace Atlas Model. This developmental process is also the story of advancement in Chinese ceramics authentication. *Survey of Ceramics Trace Model Study* marks the move into a new chapter of this story.

Qing Dynasty and earlier, marked the first period; Republic Era until now, characterized by the Double-Evidence Method of Mr. Chen Wanli as well as the rise of Technological Authentication, is considered as the second age in Chinese ceramic authentication; whilst the emergence of Ceramics Trace Model Study, the third judge in determining real from fake, has brought a new revitalized strength into the world of Chinese ceramic authentication.

# 第二章
# 陶瓷迹型学的宗旨、方法与意义

**阅读提示：**

　　陶瓷迹型学的诞生是科学引领，社会文明发展的必然产物。陶瓷迹型学是新诞生的交叉性学科。本章重点说明：陶瓷迹型学开辟了陶瓷鉴定的新领域，是一种创新的陶瓷鉴定方法，是对传统鉴定方法的继承、发展及优化，它适合于学制教育的统一传播模式。

**主要论述问题：**

1. 陶瓷迹型学的宗旨是什么？
2. 陶瓷迹型研究初级阶段的方法是什么？
3. 陶瓷迹型学有什么意义？

## Chapter 2
# The Purpose, Method, and Significance of Ceramics Trace Model Study

**Text Note:**

Ceramics Trace Model Study was born out of the earnest push of Science; the inevitable outcome of society's development towards higher civilization, it marks the new emergence of an interdisciplinary field of science. The main purpose of this chapter is to explain: How Ceramics Trace Model Study has opened a new realm of ceramics authentication; why it is a creative method of ceramics authentication; how it is the expansion, development, and optimization of the traditional intuition model approach to authentication; and why it suits the Higher Education propagation system.

# 第一节
# 陶瓷迹型学研究宗旨与方法

## 一、陶瓷迹型学研究宗旨

陶瓷迹型学研究宗旨是：突破传统陶瓷鉴定以眼学、人证的模式断定陶瓷真伪及年代的方式，将其发展为借助科学仪器，在显微及微观结构中，断定陶瓷真伪及年代。在研究的初级阶段，主要以比对陶瓷迹型标准图谱模型为主，辅之以对迹型现象及其规律的逻辑论证的物证模式，断定陶瓷真伪、年代及窑口；在研究的高级阶段，以迹型自证的模式，断定陶瓷真伪、年代及窑口。

## 二、陶瓷迹型学研究方法

陶瓷迹型学的研究方法属于科学方法。

韦氏大辞典中对科学方法是这样定义的："科学方法是一种有系统地寻求知识的程序，涉及了以下三个步骤：问题认知与表述；实验数据收集；假说构成与测试。"[15]

科学方法是在辩证逻辑和形式逻辑思路的指引下，在观察自然现象，获取新知识，或修正与整合先前已得的知识过程中，所使用的一套程序和手段系统。科学方法是人类所有认识方法中比较高级和复杂的一种方法。有效的科学方法能提高科学研究的效率和增强科学研究的效果，但它不是省略勤奋求速胜的秘方。

科学研究方法与其他获得知识的方法的不同主要在于，科学方法预测的理论依赖于事实的证明：当理论的预测被事实肯定时，理论成立；当理论的预测被否定时，理论被质疑。

### （一）陶瓷迹型研究的程序系统

#### 1. 观察步骤

将陶瓷釉层作为科学认识的主体，观察陶瓷釉层，发现问题、记录问题并定义问题。

#### 2. 搜集步骤

全面搜集和问题有关的标本与数据。

#### 3. 解说步骤

对观察、定义的问题加以解释，抓住主要矛盾，剔除非本质方面。

#### 4. 评估步骤

分析搜集的标本与数据，归纳与概括，形成科学假说。

#### 5. 实验步骤

设定假说之后，可着手进行相关实验，通过进一步的观察与试验去证实预测的结果，进行评估。

#### 6. 公开结果

将研究结果充分公开，让更多的人参加观察与检验，获得统计学的可信度。

### （二）陶瓷迹型研究的手段

#### 1. 初级阶段的研究手段

（1）标本选取

标本选取是进入观察程序的前提与开始。标本选取的正确与否，直接关系到观察结论的正确与否。选择正确真实的标本，是科学研究的充分必要条件。因此，被选作观察研究的标本，必须是有明确发掘时间、发掘地点及年代依据或有明确传世记载的陶瓷整器或残片。

（2）观察工具

科学研究简单地说就是，观察事实，整理事实，从中发现规律，作出结论。使用观察工具是获取科学知识的重要手段。

陶瓷迹型的观察工具，如果是专业研究，可采用500－600倍（或

# Section 1:
# The Purpose and Method of Ceramics Trace Model Study

## 1. The Purpose of Ceramics Trace Model Study

The purpose is to break through the traditional human witness based intuition model form of dating and appraisal through the use of scientific instruments within the microscopic and microcosmic structure. Within the entry level, the goal is to progress towards having a comparative atlas model and material evidence based trace phenomenon laws and logical theories for determining production date, kiln and authenticity. Within the advanced level, however, the goal is to achieve a self-supporting proof for determining production date, kiln and authenticity through a piece's own traces.

## 2. The Method of Ceramics Trace Model Study

The Ceramics Trace Model Study research method utilizes the Scientific Method.

Merriam-Webster's definition of the Scientific Method is: "Principles and procedures for the systematic pursuit of knowledge involving the recognition and formulation of a problem, the collection of data through observation and experiment, and the formulation and testing of hypotheses." [15]

The Scientific Method, under the direction of dialectical logic and formal logical thinking, is the whole set of procedures and methodologies used in the obtaining of new knowledge or the amendment and integration of previously obtained knowledge about the natural world. The Scientific Method is one of the highest levels and complexity of all human knowledge methods. Effective, the Scientific Methods is able to heighten our scientific research efficiency and production, but is not a secret recipe for quick results or justification for omitting diligence.

The Scientific Method is primarily different from other knowledge pursuits in that, the Scientific Method is focused on the measurement of theories through concrete reality. When the theory has been substantiated through tested fact, the theory is accepted; but when a theory is demonstrated to be false, the theory is questioned.

### A. The Ceramics Trace Model Study Procedural System

I. Observation Process

With the ceramic glaze as the main scientific focus; observe the glaze level, discover queries, record queries, define queries.

II. Collection Process

Comprehensively collect samples and data relating to queries.

III. Illustration Process

Give clarification of the observed and defined queries, weighing chief contradictions and dismissing non-essential facets.

IV. Appraisal Process

Analyze the collected samples and data, categorize and summarize, form scientific hypothesis.

V. Experiment Process

After forming Hypothesis, begin the relative experimentation.

以上）数码光学显微镜；如果是收藏爱好者，可选用150倍光学显微镜，也能达到较好的观察效果。

（3）观察方法

目前观察陶瓷迹型最常用的技术是使用数码光学显微镜，将数码摄像头垂直紧贴陶瓷釉表面，缓慢在釉表面移动，然后通过电脑屏幕观察陶瓷釉层的显微结构。但这种技术只能够观察釉层的二维形貌。

在研究过程中，广州东方博物馆美籍科研人员马士宁先生，发明了观察釉层三维形貌的观察方法。我们将这种观察方法命名为"马士宁三维成像法"。

通过观察研究釉层的三维形貌，能够有效厘清不同迹型类别的风化特征与机理特征，为各种陶瓷迹型类别风化的质变与量变特征提供有利证据，并且能帮助我们分析、解释这些特征的形成。

其次，在观察陶瓷标本时，要求全方位的观察。无论是陶瓷整器或残片，观察时要对陶瓷釉层整个面积进行全方位扫描式观察，尤其是陶瓷整器，观察要从口沿、器身至底部，不要遗漏，同时要做好观察不同部位的标记记录。

因为，只有全方位扫描观察，才能发现更多的迹型证据。

只有全方位扫描观察，才能对迹型进行量化统计。

只有全方位扫描观察，才能确保不错过更接近标准图谱模型的迹型。

只有全方位扫描观察，才能在同一器物上发现迹型的多样性和复杂性特征。

（4）分类整理

将观察到的所有现象归纳整理，将典型的迹型现象分类，从中发现各迹型类别自身发展变化的规律以及各类别之间关系的规律。

在对陶瓷迹型的分类整理中，主要运用归纳的方法分类；在对瓷迹型类别的研究中，对迹型类别自身发展变化的规律和迹型类别之间关系的规律研究，主要运用类比与演绎的方法。

## 2. 高级阶段的研究设想

古陶瓷釉层风化的过程实质上是风化产物在釉层的物理堆积和釉层的化学溶蚀，它还涉及到釉层微观结构（可到分子、原子级）的变化。研究这一过程，既可以运用比初级阶段更高级的科学仪器，采集釉层迹型数据，建成数据库，用于比对鉴定，这是物证模式的升级；还可以利用计算机扫描分析釉层风化堆积物的排列方式及化学成分、釉层溶蚀后的形貌及釉层微观结构的变化，直接做出判断，这是陶瓷迹型的自证模式。

Through further observation and testing, work towards proving the predicted results and carry out appraisal.

### VI. Publish Results

Fully disclose research results publically, allow for additional parties to conduct observation and experimentation, and obtain statistical data on the level of reliability.

## B. The Ceramics Trace Model Study Approach

### I. Entry Level Research Approach

#### a. Sample Selection

Sample selection is the premise and initial step of the observation process. Proper sample selection directly correlates to whether or not the proper observation conclusion can be made, therefore selecting precise and accurate samples is imperative for scientific research. Because of this, it is imperative that the members of the sample selection are ceramic wares or shards of definite excavation record and age, or possess concrete provenance.

#### b. Observation Tools

Science is the survey and arrangement of facts, and from these, deriving laws and putting forth conclusions. The use of observation tools to further our scientific knowledge is an instrumental part of the approach.

For professional trace model research of ceramics the possession of a 500-600X (or higher) magnification digital microscope camera is necessary. However, for the hobby antique collector, a 150X magnification hand-held optical microscope will allow for reaching relatively sufficient observation results.

#### c. Observation Technique

Currently, the most common observation tactic is to slowly scan the digital microscope camera across the glaze face of the ceramic piece while observing the microstructure of the glaze level on the corresponding computer screen (via USB connection). Note: This method will only offer the viewer a two-dimensional planar view of the glaze level.

However during our research, we have also developed a methodology for three-dimensional observation of the glaze surface, called the Bunney Method.

Research using this observation method has allowed us to effectively clarify the weathering process and mechanical structure of several different types of ceramic traces. It has helped to provide advantageous evidence on the characteristics of qualitative and quantitative change of several ceramic trace types, and to analytically interpret the formation of these characteristics.

All-inclusive observation is the primary rule and requirement for ceramics observation, regardless of whether it is a ceramic shard or a whole piece. When observing a ceramic piece it is important to scan the entire object. This is specifically true of whole piece ceramics. When scanning, it is critical to start from the mouth of the vessel and work down the body to the foot of the piece leaving no section without being scanned; while being sure to leave corresponding markings to record what areas have been covered or areas of important findings.

It is only through all-inclusive scanning observation:

More trace characteristics can be discovered; traces can enter fully quantitative statistical recording;

we ensure the most precise Standard Trace Atlas Model comparison; and

on one piece we can observe the phenomenon of trace multiplicity and complexity.

#### d. Classification Arrangement

All observable ceramic traces need to be summed up and classified into the characteristic categories. From these trace categories; we must discover the regular patterns of changes within each trace as well as their relationship with other trace categories.

Ceramics Trace Model Study categorization primarily utilizes an induction method to arrange categories. The research of the patterns taking place within each trace category and the laws that govern the inter-trace category relationships primarily utilizes an analogy and deduction method for further classification into types.

## II. Advanced Level Research Approach Concept

The weathering process of the glaze level is essentially the physical accumulation and the chemical corrosion of the sedentary byproducts; which can include the microcosmic structural changes of the glaze level (on a molecular or even atomic level). In order to research these processes, it is possible to utilize apparatuses of a higher technical capability than that of the entry level in order to record the glaze level trace data and establish a database for use in comparative authentication, which is an upgrade in the evidence model. Further, we can use computer analysis to directly determine a piece's authenticity by scanning the sedentary byproduct arrangement and chemical composition as well as the appearance and microcosmic structural changes caused by glaze level corrosion. This is a example of ceramics trace self-supporting proof.

## 第二节 陶瓷迹型学研究意义

### 一、建构陶瓷迹型标准图谱模型

爱因斯坦说过:"真科学的精髓在于将事物尽可能简化,但又不过分简化。"这是建构迹型标准图谱模型的基本原则。陶瓷迹型研究目的之一,是要建构迹型标准图谱模型。

陶瓷迹型在绝对风化时间内,不同的风化强度不仅能够反映陶瓷自身经历的从某个朝代至今的时间过程,还能反映它们存放环境的变化状况。不同时期不同类型不同存放环境的陶瓷迹型存在各自的普遍性特征,这些在绝对风化时间内形成的迹型普遍性特征,可看作是某个时期某种类型陶瓷迹型的代表特征。

所以,只要明确某个年代某个种类陶瓷迹型风化强度的普遍特征,以及其在不同存放环境中生成、发展的变化规律,并以此为参照系,就可以推理识别相同年代、甚至不同年代的陶瓷。

陶瓷迹型标准图谱模型就是对某个年代某个种类陶瓷迹型风化强度的普遍特征的归纳总结,也是对陶瓷迹型本质及规律的高度概括,同时反映了迹型的量变特征和迹型变化的时间过程性。它既可解释陶瓷迹型的本质现象,又是陶瓷鉴定真伪所需的比对物证系统。它类似科学研究中的其他模型,比如经济学模型,人们借助模型可深入研究,从而发现和解释经济规律;再如机械运动模型,三大力学定律的抽象,便于人们理解和揭示机械运动的规律。

需要指出的是,建构陶瓷迹型标准图谱模型,一组是不够的,因为要断定某个历史时期的陶瓷,会涉及到不同窑口、不同用料和不同工艺。不同窑口甚至同一窑口在不同时间内,陶瓷原料的配方可能是有变化的。这些情况都会造成不同的迹型现象。

所以,在建构陶瓷迹型标准图谱模型时,可根据釉的不同配方及烧制工艺,分门别类,建构多组图谱模型。

### 二、标准图谱模型比对鉴定易于大众掌握

眼学传授,若遇见悟性不高的弟子,在说不清道不明时,只能以"难以言传,只可意会"作结语。所以拜师易,学会难!纵是当代的大师,自信度也不高,中国国家博物馆原研究馆员姚青芳说:"我曾经问过史树青先生,您的准确率能有多少?他自己说,大概百分之五十吧。我还有个朋友问过耿宝昌他看东西的准确率,他的回答是,我比一般人能高些,差不多有个七成吧。这当然是两位老先生相对谦虚的说法,但反映出一个问题:'眼学'的准确率究竟能有多高?"[16]

国家文物鉴定委员会原秘书长刘东瑞说:"我们知道仪器鉴定主要分两大类,一类是测年份,一类是测成分。这两类方法都需要和古代的样本进行对比。据我所知,国内现在并没有建立起一个权威的数据库。"[17]再说,机器交给你的鉴定结论,如"器身白釉成分与元末明初时期青花瓷釉成分符合","成分符合"并不说明必然到代,这并没有回答器物本身确凿真伪和确切年代的问题。

《陶瓷迹型学概论》研究起来,门坎较高,挑战重重,但它创新和扩大了鉴定领域,鉴定以图比对,图文互证,迹型物证指向性清晰,事实胜于雄辩。

显微观察发展、延伸了眼学功能。比对迹型标准图谱模型可以看得更真切。学懂了,易于上手鉴定,能很快提高识别假货的能力。比对迹型标准图谱模型,靠谱的,再辅之逻辑证明,可判为真;不靠谱的,有的可断为假,有的则需要进一步去研究。因此,《陶瓷迹型学概论》是普及陶瓷鉴定知识的启蒙篇章。社会需要无数的新篇章续写,到那时,陶瓷鉴定将会走下神坛,走向科学,走到老百姓中间。

# Section 2:
# The Significance of Ceramics Trace Model Study

## 1. Construct Ceramic Standard Trace Atlas Model

Einstein said, "The true essence of Science is to simplify things, but not over simplify them." This is the guiding principle in constructing a Standard Trace Atlas Model. The purpose of Ceramics Trace Model Study is to construct a series of atlas models.

Different weathering severities of ceramic traces within the absolute weathering time not only can reflect the passage of time of the ceramic piece, but also reflect the change of its post-firing environment. The different post-firing environments of each type of ceramic within each dynasty has specific typical characteristics; these characteristics formed within the absolute weathering time can be used as representative characteristics of each certain type of ceramics within that dynasty.

Therefore, as long as we can define the typical characteristics of weathering severity of a certain type of ceramic within a certain dynasty, as well as the laws that govern the trace change emergence and development within different post-firing environments; we can ascertain knowledge to help distinguish between similar and dissimilar ceramics.

A Standard Trace Atlas Model is the summarization of the representative characteristics of each type of ceramic within each dynasty as well as a summarization of the ceramic trace essences and laws, and proves the quantitative change characteristics and the passage of time inducing trace changes. They not only help to expound upon the intrinsic qualities of ceramic traces, but also act as the comparative material evidence needed for ceramic authentication. They are similar to other scientific models, such as economics models, in that people can use these models to further deepen their research to discover and explain the laws of economics. Further, they resemble the mechanical movement model, in that people can use them to not only explain, but to illustrate the laws of mechanical motion.

It is important to point out that the construction of a single atlas model is not enough. This is due to the fact that in authentication of a certain period of ceramics, it is essential to touch on different kilns, different materials, and even different techniques used during that time; this is because within one kiln during different periods, raw material recipes went through vast changes. These variables can all create different ceramic trace phenomenon.

Therefore, when constructing atlas models we must classify on the basis of different glaze recipes; constructing several different atlas models. However, the fewer models the cleaner, with just enough to fully grasp the representative characteristics.

## 2. Atlas Model Comparison Easier for Layman to Grasp

If the propagation of the intuition-based method meets with a pupil that finds it hard to comprehend a certain theory, the master would say, "It's hard to explain in words, it can only be perceived by intuition." It is easy to become an apprentice, but very difficult to master. In fact, it was the evident lack of self-confidence among China's top masters that led Yao Qingfang, research fellow at the National Museum Research Institute, to surmise: "I once asked Mr. Shi Shuqing: 'What is your accuracy rate?' He returned, "Oh about fifty percent." I have another friend who asked Mr. Geng Baochang the same, he answered, "I am probably a bit higher than the rest, probably seventy percent." This is obviously spoken

正如德国大哲学家康德在倡导启蒙运动时所强调的，人类要摆脱他们所加之于其自身的不成熟状态。要摆脱在陶瓷鉴定领域的不成熟状态，就要通过学习科学，从愚昧走向智慧，并勇于运用智慧。

## 三、从听鉴定结论到看鉴定结论

眼学方法鉴定，是把鉴定结果说给你听，是要你听明白。然而，当多位鉴定专家鉴定陶瓷真伪时，即便他们都断定其为"真品"，但表达"真"的理由，往往不尽相同；表述逻辑，也因人而异。于是，对于听者，明白的程度是不一致的，有些人明白的多一点，有些人则少一点，有些人甚至听不明白。如果他们断定文物为假，部分专家说出的论据，又不那么令人信服。

建立在述说形式上的鉴定结论，如果专家双方表述观点对立，当分不清真理在谁手中时，争论往往会不了了之。正因为这种不了了之、说了都不算的现象，就给目前真假混乱的陶瓷市场，添了更多混乱。

百闻不如一见，用迹型标准图谱模型比对的鉴定方法是将鉴定为"真品"的理由，让你观看，让你比对，同时将迹型产生的原因、机理以及迹型之间关系的规律，以逻辑论证的形式阐述出来，从而分辨被鉴定器物的真伪。

例如，在鉴定一件元青花瓷器时，当专家之间发生争执时，人证不可靠了，这时候迹型标准图谱模型就是可靠的物证系统，用它来比对该瓷器的"DNA"是否靠谱，让事实说话，眼见为实，再用逻辑论证诸多迹型以及它们之间的关系特征是否是长期自然风化形成的，从而辨别其真伪。

迹型标准图谱模型的客观事实就像网球比赛中的"鹰眼"，让争执双方无话可说，因为事实胜于雄辩，因为靠"谱"，因为得到了事实与逻辑的证明。进一步来说，还可以将鉴定结果用数字表达出来，让你在数字表达中，进一步认知所鉴定器物趋真或趋假的识别程度。数字所表达出来的知识体系，在反复验证的过程中，追求的是逐渐接近真实的客观判定，而不是任何权威。

## 四、从鉴定人证到鉴定物证

眼学方法对陶瓷进行真伪、年代及窑口的鉴定，形式上是鉴定人的一种表述，这种表述的结论就是人证——口说为凭。但由于缺乏物证，自己又不能圆满证明自己所说的，所以鉴定结论往往是或然的。

我国的刑事诉讼法修正案在2013年开始实施。"向科技要警力"的口号不断得到落实，不断向"从案到人，口供为王"的传统侦察观念发出挑战。

"由人到案，证据定案"，是物证时代科学的侦察方式。

如果对涉及古陶瓷刑事案件的涉案器物进行司法鉴定，多个专家对同一件器物的鉴定出现不同的结论时，就无法为侦察机关提供可靠而科学的鉴定结论。

在涉及古陶瓷的民事案件中，原告与被告方各自聘请专家对器物进行鉴定，如鉴定有矛盾，双方不服，法院只能再次请来指定专家。如果鉴定意见再不能统一，又不能提供科学、准确且客观的司法鉴定证据，那么案件就会拖而不决，审而不服，判而不公。

陶瓷迹型学，是从陶瓷自然风化的迹型中，总结提炼出迹型标准图谱模型，图谱模型就是物证。这种中立、公正、客观、量化的直观物证，可望在司法鉴定工作中运用于法庭，可填补我国文物物证司法鉴定的空白，让文物鉴定从权威说了算的时代走向物证说了算、科学说了算的时代。

## 五、陶瓷迹型学是陶瓷鉴定方法的一种补充

当今中国的陶瓷鉴定，还是以眼学鉴定为主。

中国古陶瓷的眼学鉴定，特别类似于中国的中医学。中医看病，以脏象为核心，望、闻、问、切，强调礼、法、方、药的有机统一。中医把人看成与自然不可分割的整体，强调整体观。人身之阴阳，顺天地之规律，达到平衡则无恙。

陶瓷眼学鉴定，看胎、抚釉、释纹饰、断工艺，再联系那个时代的文化背景，然后断代。

然而，中医的困惑是难看急诊，难治结核，难于进入微观。进不了微观，就难于量化研究。无量化研究的学科，就有伪科学之嫌。我国的中医学，不属于实证医学，是非西医边缘化的补充疗法，因此，一直得不到许多西方国家的官方认可。

在中国，西医介入了中医研究。例如，西医发现，中医所谓的经络，既不是神经也不是血管。它

out of the humility of two great masters, but it does beg the question, "How precise can intuition method's correctness actually be?" [16]

The former National Cultural Relic Authentication Committee's Secretary General, Liu Dongrui, once said, "We know that apparatus authentication has two main categories: One for surveying age, and the other for surveying composition; both of which need to be carried out in comparison with archeological specimen. As far as I know, we [China] have yet to construct an authoritative database." [17] Moreover, even the conclusion given by such an apparatus would look some thing like this: "The composition of the white glaze is consistent with late Yuan early Ming Dynasty Underglaze Blue Porcelain glaze composition". It does not answer the question of the piece's authenticity or its production date, and lacks the ability to leave the individual with peace of mind.

Even though the introductory threshold of this survey and entry-level research is full of challenges; the text is easy to comprehend, with pictures for comparison, and the material trace evidence is direct, allowing facts to speak for themselves.

Microscopic observation develops and expands intuition-based method's function, allowing you to see more clearly. Once mastered, it will decrease the difficulty in authentication and heighten your ability to distinguish real from fake. When using the Standard Trace Atlas Model for comparison, those that match within logical verification limits can be determined as genuine, and of those that do not match, while some can be ruled as fake, the others require further research. *Survey of Ceramics Trace Model Study* is the entry-level textbook for imparting ceramics authentication knowledge; innumerable related texts are just waiting to be written. It is through this process that ceramics authentication can be taken down from its mysterious pedestal, as we move towards a more systematic, scientific, and user-friendly playing field.

It is like the Enlightenment, initiated by German philosopher Immanuel Kant, who emphasized: "Man must emerge from his self-incurred immaturities." We must emerge from the immaturities in the realm of ceramic authentication, following the lead of others that have gone before, we must move from ignorance towards wisdom, and greater still, we must have the bravery to use that wisdom.

## 3. From Hearing Appraisal to Seeing Authentication

The intuition-based naked-eye appraisal method is to tell the listener the results of the appraisal, and it is up to the listener to understand. When several experts authenticate the same piece, unanimously concluding that a piece is 'true', they may still offer different explanations as to the validity and basis of their conclusions. Additionally, the stated logic used in the judgment of a 'true article' varies with each individual. Therefore, listener understanding is inconsistent; some able to comprehend more, some less, even to the extent that some don't grasp it at all. If a relic is judged to be 'fake', the argument given usually does not leave the listener feeling entirely convinced of its veracity.

In this verbal form of appraisal, when two experts come to an impasse or have conflicting viewpoints as to the true nature of a piece, the dispute is often settled by agreeing to disagree. This 'agree to disagree' practice is precisely what is creating more confusion in an already chaotic ceramics market.

It is better to see for yourself than to hear from others. The grounds for authenticating a 'genuine article' by using the Standard Trace Atlas Model comparative authentication method allows the onlooker to observe and compare for himself. At the same time, the causes of trace formation, trace mechanisms, and inter-trace relationship laws are elaborated via logical proofs, allowing for the judgment of the item's veracity.

This is to say, when looking at the Yuan Dynasty Underglaze Blue 'family' of porcelain; legitimacy is gained by visually comparing its DNA against the mapped genome of that family, letting truth speak for itself, and letting the observer to see the truth for himself. When there are discrepancies among experts, human witness is no longer reliable; we need to depend on material evidence provided by a Standard Trace Atlas Model, utilizing logic to prove that the traces and their inter-trace relationships came about through natural weathering and are currently infeasible to forge.

The Standard Trace Atlas Model's objective precision is similar to that of 'Hawkeye' instant replay in tennis and other

的物质特征究竟是什么？西医的解剖学没找到，用声、光、电、同位素还是没找到。通过一些科学家的不断努力，建立了经络的科学假设——"低流阻通道"。这个假设认为，经络是流动的液体，像大海似的无管道流动，由液体系统组成的经络运行在人体中。中医的经络神秘性，还是得到了有物质对应的西医科学假设的解释。

中医为了发展，还是引进了西医，而且被西医纳入了微观研究。而今，中西医结合的治疗方法已被广泛采纳。同理，陶瓷研究进入显微及微观世界，进行量化分析，并非否定眼学，而是在眼学基础上，辅之证明、证据，从而使陶瓷鉴定更加科学，更加准确，更加有说服力。

如果说，眼学鉴定方法有赖于"标型学"原理，那么，迹型标准图谱模型比对鉴定就是依据"迹型学"原理。

科技鉴定中最广泛使用的元素分析法，只是对陶瓷的成分进行检测，不进行断代说明，而且这种方法仅可以排除部分假冒产品，其准确性还有赖于数据库的继续完善。

陶瓷迹型鉴定，是对眼学鉴定的有力补充，是眼学鉴定、科技仪器鉴定的发展与延伸。如果综合热释光、元素分析、标型特征、迹型特征等去综合判断一件陶瓷，其结果就更接近真实。正如朱清时院士所说："在特定的情况下，科学跟着经验走，但是，最终经验要跟着科学走。"[18]

## 六、陶瓷迹型学是历史发展的必然趋势

科学的发明，技术的进步，让人类不断对客观世界有了更深刻的认识。在认识过程中，人类不断碰到问题、解决问题，掌握推动世界进步的主动权。文明战胜愚昧，科学战胜迷信，这是历史发展的必然趋势。

但是，在这一过程中，任何创新、革新或"跨越鸿沟"，都不是一帆风顺的。数百年前，仅仅"地心说"、"日心说"之争，就让意大利科学家布鲁诺付出了生命的代价。

百年企业柯达公司的轰然倒塌也说明了这一点。随着数码科技的崛起，作为传统摄影使用的胶卷主要生产商之一的柯达公司，终于在2012年宣布破产。具有讽刺意味的是，柯达早在1975年就发明了数码相机，并拥有多项数码摄影专利。由于直到20世纪90年代末，世界上的人绝大多数还是采用胶片感光技术拍摄照片，柯达选择继续保守地在这个领域当霸主，对数码新技术选择了忽视。这家拥有130年历史的照相胶卷先驱企业，由于缺乏改革创新的动力和勇气，最终还是被这个日新月异的科技时代淘汰了[19]。

相反，与柯达竞争的富士却锐意革新，与时俱进。时至今日，柯达市值不到1.5亿美元，富士市值近120亿美元。

千百年来，中国的陶瓷鉴定，一直靠眼学，一直是师徒式的传承。但在科学技术日益发展的今天，在迫切需要社会诚信、社会管理的今天，将量化研究、显微及微观研究、团队式研究纳入高等教育体系，必将成为中国陶瓷研究与鉴定的发展趋势。

任何变革都会遭遇来自保守力量的阻力。保守力量过去存在，现在仍然存在。革新者应该从无数科学家为科学献身的精神中吸取意志与力量，更应该在柯达的轰然倒塌中吸取经验与教训。

## 七、陶瓷迹型学是人类认识发展的必然

人类对客观事物的认识，是从感性认识到理性认识，从事物现象深入到事物本质，从低级阶段发展到高级阶段的过程。这反映了人类对客观事物认识的一般规律。

眼学仰仗标型学和那个时代的社会背景对陶瓷进行断代研究，在认识论中尚属于经验型的感性认识阶段。但是，利用显微及微观技术，把陶瓷的真伪及断代判定引入显微或微观，就可以建立一个判别陶瓷器真伪及年代的科学系统，可称之为专家系统。

因为科学是不迷信大师的，科学的词汇里不存在事实以外的权威。

判别陶瓷年代及真伪问题，实质上就是模式识别的问题，因为任何判别的问题都是模式识别的问题。

将具有可靠性的研究标本，在显微观察的基础上，将时间在釉表面及釉层中留下的风化痕迹以历史的方法展示，以逻辑的方法分类，提炼出多维向量，并加之权重，计算出分值，再根据分值大小，进行断代的可靠性评估。

这种新方法，发展了眼学鉴定中以述说特征评估的方法。新方

sports. It alleviates confrontation by allowing visual evidence, through the use of uniform standards, to override a heated argument. Using concrete formulas to express authentications allows us to further understand whether or not an appraised piece is real or fake. This knowledge system represented by mathematic formulas is gradually approaching accuracy through the pursuit of repeatable verification rather than the pursuit of any one person's authority.

## 4. From Human Witness Appraisal to Material Evidence Authentication

Intuition method is used to describe the period, kiln, and authenticity of an item. Essentially, it is the appraiser's explanation of a piece; this form of verdict through explanation is called human witness, 'My word is my proof'. However, due to a lack of material evidence, there is no way to satisfactorily testify to what they are saying, therefore appraisals can only ever remain as possible and not resolute.

2013 marked the implementation of a new amendment to the criminal justice system in China, marking the desire to move forward with the slogan "A More Technological Police Force". This change challenged their traditional investigation mentality, which saw 'testimony as king', and moved them towards a new material evidence era in which 'evidence is king'.

If during the course of a criminal proceeding, evidence is brought forth for the authentication through expert opinion, and two experts have opposing views, a purely material evidence based decisive verdict is unattainable.

If in a civil case the plaintiff and defendant bring forth evidence that requires the authentication or appraisal from their own respective experts and conflict arises, neither side is appeased. In this case, the courts must appoint an additional expert to act as judge on the authentication matter. If a unified opinion still cannot be reached, an objective verdict based on material evidence testimony cannot be reached; creating a situation of uncertainty, difficult regulation, and an unjust verdict.

Ceramics Trace Model Study draws from the natural weathering traces on ceramics to surmise and construct a Standard Trace Atlas Model, a form of comparative material evidence. This type of neutral, impartial, objective, and quantitative evidence could be utilized in the court of law and fill the current void felt by the courts in the field of cultural relic authentication. This would allow the current method, which utilizes the authority of an expert's word to authenticate, to be replaced by a new method that calls for scientific material evidence to act as the final verdict. This allows relic authentication to move from an era of 'experience is king', to an era allowing science to be the final authority.

## 5. Trace Model Study Complementary to Other Authentication Methods

Today's method for authenticating ceramics in China is still primarily intuition-based naked-eye appraisal.

Chinese intuition-based appraisal of ancient ceramics is particularly analogous to the field of Chinese Medicine. Chinese Medicine looks at a sickness for visible manifestations indicating physiological function as well as pathological changes of the internal organs. It bases examination on outward manifestations noticeable by the five senses and stresses the organic nature of the body, oneness with nature, as well as a need for balance with nature. It views the body and nature as inseparable, and approaches observation holistically.

Similarly, intuition-based appraisal utilizes looking at the clay, touching the glaze, explaining the patterns, and judging the technique in the context of the culture of that time period to reach a final conclusion. However, Chinese Medicine finds difficulty in emergency medicine, curing unseen nodules, entering into a microscopic observation method, and quantitative research. Fields without quantitative research fall under the suspicion of pseudoscience. Chinese Medicine is not considered a proof-based medicine and has been marginalized to a supplementary treatment form of Non-Western Medicine and has, therefore, not attained the approval of many Western governments.

In China, Western Medicine professionals have entered into the research of Chinese Medicine. For example, Western Medicine researchers found that the acupressure system of Chinese Medicine is neither nervous nor cardio-vascular.

法以眼见为实的展示，以逻辑的缜密证明，以更多量化特征的学理证明，将人们对陶瓷研究断代的感性认识提升到理性认识阶段，以客观的陶瓷迹型，辅之以眼学判断，最终让我们对陶瓷真伪和年代的判断更趋向真实。

在陶瓷迹型显微及微观领域中的研究，有所认识，有所发现，有所创造，才能更准确地识真驱假，传承与弘扬真正的民族文化。

Then what are its material characteristics? The researchers using sound, light, electricity, and isotopes were unable to anatomically locate this system. Through the arduous research of scientific professionals, they were able to hypothesize that this acupressure system is a "free flowing and unrestricted" liquid similar to that of an ocean without set pipelines or passages directing flow. It is through this acupressure system that this liquid flows throughout the body. It was only through this Western Medical hypothesis that the mysterious acupressure ideology was finally explained according to corresponding matter.

Chinese Medicine, for the sake of survival, has begun to recognize Western Medicine, and has thus been pulled into microscopic and microcosmic research. Similar to the way in which integrated Chinese and Western Medicine has recently become widely accepted as a treatment method; microscopic research of ceramics does not completely dismiss intuition-based appraisal, but offers supplementary evidence based on the foundation of the intuition-based method. It allows the authentication of ceramics to be more scientific, accurate, and credible.

If we are to say that intuition-based appraisal is founded on the comparative principles of "Typology", then the Atlas Model comparative method is founded on the principles of "Trace Model Study".

Microscopic research and authentication of ceramics is a powerful complement to the intuition-based method. It is the extension and expansion of intuition-based appraisal. When we can synthesize TL testing, XRF, Typology, and Trace Model Study to judge the authenticity of ceramics, then we will move even closer to absolute accuracy. It is like Zhu Qingshi said, "Under special circumstances, science follows in the wake of experience, but ultimately experience must come through science." [18]

## 6. Historical Development's Necessity for Ceramics Trace Model Study

Scientific development and technological progress continually help mankind to deepen its understanding of the objective world. Our continuous running into problems and solving problems gives us the enterprise to continually press forward. Civilization overcomes ignorance, science overtakes superstition. This has been a constant tendency throughout history.

However during this process, any trail blazing, innovation, or 'giant leaps' often come against strong winds. A few hundred years ago, merely the battle between geocentric and solar-centric theory cost some scientists their lives.

Recently, the crashing fall of the century-old Kodak Company also speaks to this; during the rise of digital technology, the primarily traditional roll film based company finally declared bankruptcy in 2012. The irony is that in 1975, Kodak was responsible for the invention of digital photography and possesses several related patents. As the majority or the general public, well into the Nineties, did not use digital photography, but rather the traditional film cameras, Kodak decided to hold off on developing digital imaging and hold their position as the 'king of film'. This 130-year-old industry giant, due to a corporate culture lacking the motivation and courage to be revolutionary, met its fate at the hands of this hi-tech age [19].

Conversely, Kodak's main competitor, Fujifilm, has remained resolute on innovation and keeping up with the times. Today, Kodak's market value is less than 150 million USD, while Fujifilm is now 12 billion.

Over the centuries, Chinese ceramic appraisal has relied on intuition-based master-apprentice tradition. However, in this high-tech age, society urgently needs the trust that quantitative, microscopic, group-based research in stride with the tertiary education system. It is the inevitable trend that ceramic authentication must follow.

Any transformation encounters hindrance from conservative forces. Conservative forces were present in the past, and are present now. Innovators must not only draw determination and strength from the fervor for science of numerous scholars which have gone before; but also draw an important lesson and experience from Kodak's avoidable fall.

# 7. Human Knowledge Development's Necessity for Ceramics Trace Model Study

Human knowledge of objective things always progresses from a rudimentary stage into a higher level; from perceptual knowledge to rational knowledge; and from the appearance of things to the essence of things.

The intuition method relies on typology as well as the relative period's societal background to assign dates and carry out research. The process of forming conclusions is experience-based and uses perceptual knowledge. Utilizing the Trace Model method to determine the age and authenticity of ceramics, however, forms conclusions based on microscopic or microcosmic observation; creating a system for drawing conclusions based on the foundation of a multi-faceted model. Only this can be called an 'Expert System'; as science does not blindly follow a master's words, nor does its vocabulary contain the authority of anything out side of fact.

Differentiating ceramics' age and authenticity is essentially a problem of model selection, because any question of differentiation is a question of the model used to distinguish it.

Having reliable research samples, founded on microscopic observation, reveals to us the traces of time and history left on the body surface, glaze surface and within the glaze level. It also allows us to make logical categorizations, extract multi-dimensional fact, add weight, assign points, and according to the higher or lower results, conduct a reliable assessment of the age and authenticity of a piece.

This new method breaks the mold of simply stating the characteristics in an assessment, as is the case with intuition-based appraisal. It is a method that reveals the quantitative characteristics used in scientific certification, while bringing ceramics research from a stage of perceptual knowledge into the realm of rational knowledge. When we use objective ceramic traces to supplement intuition-based judgment, we ultimately move towards more accurate decisions about age and authenticity.

Through the realm of microscopic and microcosmic research of ceramic traces; only with full understanding, full discovery, and full innovation can we bring about more accurate authentication, as well as the honest preservation and reverence of cultural heritage.

# 第三章
# 陶瓷迹型学

**阅读提示：**

本章涵盖了陶瓷迹型学的基础概念、鉴定依据、鉴定原理、鉴定本质以及鉴定局限性等多方面内容，着重阐述陶瓷迹型学鉴定依据和传统陶瓷鉴定依据的差别，强调陶瓷迹型学鉴定依据是对传统陶瓷鉴定依据的创新和发展。

**主要论述问题：**

1. 陶瓷迹型学鉴定依据与传统陶瓷鉴定依据有什么区别？
2. 陶瓷迹型学的鉴定原理是什么？
3. 陶瓷迹型学的鉴定本质是什么？
4. 陶瓷迹型学有什么局限性？

# Chapter 3
# Ceramics Trace Model Study

Text Note:

This chapter covers multiple topics including: Ceramics Trace Model Study foundational concepts, authentication foundations, authentication principles, nature of authentication, and limitations to authentication. It is meant to highlight the differences between the foundations of Trace Model Authentication compared to that of traditional ceramics appraisal. It emphasizes that the basis of Trace Model Authentication is a breakthrough, and is an expansion on the foundation of traditional appraisal.

# 第一节
# 陶瓷迹型学基础概念

## 一、陶瓷迹型学概念系列一

### 陶器

陶器是用河谷沉积土、普通泥土等无机物质做原料，采取手工或其他方法做成所需要的形状，经过800℃至900℃的温度焙烧，使之硬化而成的物品。

由于原料等基础因素，成品坯体未曾烧结，具有无透明性，有小孔，有吸水性的特性[20]。

### 瓷器

瓷器是以瓷土或瓷石为原料，经过配料、成型、挂釉、干燥、焙烧等工艺流程制成的器物。

瓷器的化学组成主要是氧化硅、氧化铝，并含有氧化铁、氧化钛、氧化钙、氧化镁、氧化钾、氧化钠、氧化锰等。瓷胎烧结后，质地致密，不吸水或吸水率很低，胎呈白色，胎薄者又具有高透明度和一定的机械强度，击之有清脆的铿锵声。瓷器的烧成温度必须在1200℃以上[21]。

### 陶瓷迹型

陶瓷迹型是陶瓷釉表面及釉层中的显微及微观痕迹的总和。

### 陶瓷迹型学

陶瓷迹型学是研究陶瓷釉表面及釉层中的显微、微观痕迹及痕迹变化规律的学科。

陶瓷迹型学包含了各个历史时期，陶瓷的低温釉、中温釉及高温釉的迹型情况。

### 陶瓷迹型类别

陶瓷迹型类别是陶瓷在釉表面及釉层中具有相同本质属性的显微及微观痕迹的总和。

### 陶瓷迹型种类

陶瓷迹型种类是陶瓷迹型类别中具有相同特有属性的显微及微观痕迹的总和。

### 陶瓷迹型单项

陶瓷迹型单项是陶瓷迹型种类中单个显微或微观痕迹。

### 陶瓷斑块

陶瓷斑块是陶瓷在烧制过程中和后天环境影响下形成的，以不规则块状形貌分布在釉表面及釉层中的迹型类别。

### 陶瓷纹路

陶瓷纹路是陶瓷在烧制过程中和后天环境影响下形成的，以线状、条状、网状等形貌分布在釉表面及釉层中的迹型类别。

### 陶瓷气泡

陶瓷气泡是陶瓷在烧制过程中形成的以气相形式存在于釉层中的迹型类别。

### 陶瓷迹型标准图谱模型

陶瓷迹型标准图谱模型是观察研究若干个明确发掘时间、发掘地点、年代依据及有明确传世记载的陶瓷标本后，按其迹型现象提炼概括出来的，用于陶瓷鉴定比对的以显微或微观成像图片构成的物证系统。

### 数码显微镜

数码显微镜是将看到的实物图像通过数模转换，使其成像在计算机显示屏的显微镜[22]。

### 显微成像

显微成像是指使用500X数码显微镜拍摄到的实物图像。

Section 1:
# Ceramics Trace Model Study Foundational Concepts

## 1. Ceramics Trace Model Study Concepts Series 1

### • Earthenware

Utilizing aqueous sediments, normal soil, and other inorganics as raw material; using hand crafting or other means to shape; fired under 800-900°C temperatures; the final product is a hardened ceramic ware.

Due to the nature of the natural materials, the body of the piece is: Not agglutinated, non-transparent, porous, and water-permeable [20].

### • Porcelain

Utilizing porcelain clay as the raw material; it is a ceramic ware that undergoes specific mixing, shaping, glazing, drying, firing and other hand crafting techniques.

The primary chemical components consist of Silicone Dioxide (Silica), Aluminum Oxide, Ferric Oxide, Titanium Oxide, Calcium Oxide, Magnesia, Potassium Oxide, Sodium Oxide, Manganese Oxide and other oxides. After agglutination, the body is dense with little or no water permeability. The body is white with a slight translucency in thin wares. It contains a certain mechanical strength and has an audible resonance when struck. The body is translucent, glass-like, and impermeable. The body must undergo firing of temperatures above 1200°C [21].

### • Ceramic Traces

The sum of the microscopic and microcosmic traces on the glaze surface and within the glaze level of ceramics.

### • Ceramics Trace Model Study

A branch of research dedicated to the study of the microscopic and microcosmic traces and trace change patterns on the glaze surface and within the glaze level of ceramics.

Ceramics Trace Model Study encompasses the trace characteristics of Low, Medium, and High temperature glaze ceramics from each historical period.

### • Ceramic Trace Category

The grouping of microscopic and microcosmic traces on the glaze surface and within the glaze level of ceramics that possess similar intrinsic characteristics.

### • Ceramic Trace Type

The grouping within a Ceramic Trace Category of microscopic and microcosmic traces on the glaze surface and within the glaze level of ceramics that possess similar characteristic attributes.

### • Individual Ceramic Trace

Within a Ceramic Trace Type, a single microscopic or microcosmic trace on the glaze surface or within the glaze level of ceramics.

### • Ceramic Mottles

Irregular blotch shapes spread across the glaze surface or throughout the glaze level of ceramics incurred during the firing process or under the influences of the post-firing

## 单位图片

单位图片是本书中所有对应釉表面面积为0.8mm²（1.0mm×0.8mm）的显微成像照片。

## 马士宁三维成像法

马士宁三维成像法是广州东方博物馆一名美国籍研究人员马士宁发明的陶瓷釉表面三维成像的拍摄法。

陶瓷釉表面的三维成像法，属于观察方法的创新。

如图3-1-1为陶瓷釉表面的二维图片，图3-1-2为同一位置的三维图片。

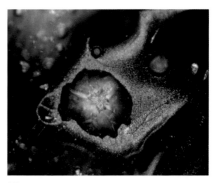

图3-1-1

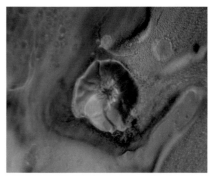

图3-1-2

## 眼学

眼学是利用已确定为某个历史时期的器物作为标准器，用于比对鉴定的学问，也可以称为目鉴学。

## 二、陶瓷迹型学概念系列二

### 青花瓷

青花瓷是在瓷器坯体上用以钴土矿为原料制成的青料描绘纹饰，再罩上一层透明釉，经高温（1280—1350℃）还原焰一次烧成的釉下彩瓷器[23]。

### 胎

胎是以粘土为主，并添加其他溶剂原料塑成一定形状，在高温作用下形成陶瓷坯体的坯料[24]。

### 釉

釉是施于陶瓷坯体表面的利用天然矿物原料按比例配合，在高温作用下熔融而覆盖在坯体表面的玻璃质层[25]。

### 青料

青料就是加工后用来绘制青花瓷纹饰的一种钴土矿物[26]。

### 着色剂

着色剂是在陶瓷的胎、釉中起呈色作用的物质[27]。

### 瓷石

瓷石是一种主要由石英、绢云母组成，含少量长石、高岭土等的矿物质[28]。

### 石英

石英是主要成分为二氧化硅（$SiO_2$），同时含有少量长石、云母、金红石等矿物的矿物质[29]。

### 草木灰

草木灰是草本和木本植物燃烧后的残余物[30]。

### 釉灰

釉灰是草木灰与石灰石的煅烧物[31]。

### 石末

石末是石英粉末[32]。

### 铅粉

铅粉是用铅加工而成的，主要是碱式碳酸铅〔分子式为$2PbCO_3 \cdot Pb(OH)_2$〕的化学组成不固定的混合物[33]。

### 硝

硝是一种主要化学成分是$KNO_3$的天然原料[34]。

### 硅酸盐

硅酸盐是由硅（Si）、氧（O）与金属（主要是Al、Fe、Ca、Mg、K、Na等）或有机基团结合而成的化合物的总称[35]。

environments.

Ceramic Mottles is a Ceramic Trace Category.

### • Ceramic Striae

Linear, stripe, or web shaped appearances spread across the glaze surface or throughout the glaze level of ceramics incurred during the firing process or under the influences of the post-firing environments.

Ceramic Striae is a Ceramic Trace Category.

### • Ceramic Air Bubbles

Gaseous pockets spread throughout the glaze level of ceramics formed during the firing process.

Ceramic Air Bubbles is a Ceramic Trace Category.

### • Ceramic Standard Trace Atlas Model

A system based on a collection of microscopic or microcosmic images taken from ceramic pieces or shards of determined excavation record and age or concrete provenance that act as a standard of material evidence for the comparative authentication of same-type ceramics.

### • Digital Microscope

An instrument for the digitizing of the microscopic images of a specific object, for the purpose of viewing on a computer screen [22].

### • Micro Imaging

Refers to the 500X digital magnification images which are taken with a digital microscope.

### • Unit Image

The 0.8 mm² (1.0mm x 0.8mm) glaze surface images used within this book; taken with a 500X magnification digital microscope.

### • Bunney Method

Matthew Bunney, a research professional at Guangzhou Oriental Museum, developed an innovative method of observation that allows for three-dimensional imaging of the glaze surface; considered a trail-blazing approach to ceramic inspection.

See the examples: Picture 3-1-1 is the two-dimensional, and picture 3-1-2 is the three-dimensional view of the same spot on the same piece.

### • Intuition Model Study

Also called Naked Eye Appraisal, it is the use of an article determined to be of a certain time period as the standard for comparative authentication.

## 2. Ceramics Trace Model Study Concepts Series 2

### • Underglaze Blue Porcelain

Also called Blue and White; refers to a type of color-under-the-glaze ceramic that uses cobalt ore as the raw material in the dye for pattern depiction onto the plain clay body, with a transparent glaze then applied over the top and fired once at 1280-1350°C [23].

### • Body

A primarily clay based material with added solvents, formed into a shape, and fired under high temperatures to become the ceramic base material. Also referred to as biscuit [24].

### • Glaze

A specifically mixed set of natural and mineral raw materials for application to the surface of a ceramic body that when fired, melts and then cools to become a glass phase layer over the body [25].

## 玻璃

玻璃是熔融后冷却至固态的非结晶（在特定条件下也可能成为晶态）无机物[36]。

## 硅酸盐玻璃

硅酸盐玻璃是主要成分为硅酸盐的玻璃。

## 釉层

釉层是由玻璃相、气相、晶相及少量原料残留颗粒构成的覆盖在陶瓷坯体上的玻璃体[37]。

## 釉层形貌

釉层形貌是釉层在一定观察角度下的三维可视状况。

## 釉层结构缺陷

釉层结构缺陷通常指陶瓷釉层分相处、相与相之间结合的结构不稳定处或釉层有孔隙的地方。

## 宏观结构

宏观结构是肉眼可观察的，不涉及分子、原子、电子等内部结构或机制的结构[38]。

## 显微结构

显微结构是指在光学显微镜下观察到的，其最小结构尺度大于晶胞（晶体中的最小单位）尺寸的结构[39]。

## 微观结构

微观结构是观察结构尺度小于晶胞尺寸的结构[40]。

## 相

相是指物理和化学性质完全相同且成分相同的均匀物质的聚集态[41]。

## 分相

分相是陶瓷釉层从熔融态冷却下来时，釉层内部质点迁移，某些组分发生偏聚，形成化学组成不同的两个相的过程[42]。

## 玻璃相

玻璃相是在陶瓷高温烧制时各组成物质产生一系列物理变化、化学反应后冷却凝固形成的一种非晶态物质[43]。

## 气相

气相是存在于烧制好的陶瓷釉层中的气体。

## 晶相

晶相是陶瓷釉表面或釉层中析出的有一定形状和颜色的结晶体。

## 晶体

晶体是内部质点在三维空间呈周期性重复排列的固体[44]。

## 析晶

析晶是从熔体或玻璃体中析出晶体的过程[45]。

## 结晶

结晶是物质在一定的物理化学条件下从各种状态转变为晶体的过程[46]。

## 润湿

润湿是液体物质附着铺展于固体表面以降低固体的表面能的现象[47]。

## 风化作用

风化作用是陶瓷釉层与周围环境中的各种物质相互作用发生物理变化和化学反应的全过程[48]。

## 风化作用强度

风化作用强度是外界环境对陶瓷釉层破坏的速度。

## 风化强度

风化强度是陶瓷釉层风化结果的高低程度。

## 绝对风化时间

绝对风化时间是指陶瓷从烧制完毕出窑后的时间延续到现在的时间过程。

• **Cobalt Dye**

Ore, processed to be used as the decoration material applied to Underglaze Blue (Blue and White) Porcelain [26].

• **Coloring Agent**

A material used to assume color on the body or glaze of ceramics [27].

• **Porcelain Clay**

Mineral composite primarily composed of quartz and mica as well as a small portion of feldspar, kaolin, or other trace elements [28].

• **Quartz**

Mineral composed primarily of Silicone Dioxide ($SiO_2$) with other traces of feldspar, mica, rutile, and so forth [29].

• **Plant Ash**

The byproduct of burning grass, timber, and other flora [30].

• **Glaze Ash**

A calcite material composed of plant ash and limestone [31].

• **Quartz Powder**

Quartz ground to a fine powder [32].

• **Lead Powder**

An unfixed mixture of the alkaline lead carbonate; a processed form of lead, as a fine powder [33].

• **Niter**

Also known as Saltpeter, it is a type of primary chemical composition, found as the natural raw material Potassium Nitrate ($KNO_3$) [34].

• **Silicate**

The generic term for all chemical compounds formed by the combination of Silicone (Si), Oxygen (O), and a metal (primarily Al, Fe, Ca, Mg, K, Na) or organic radical [35].

• **Glass**

A non-crystalline (under certain circumstances it can crystallize) inorganic material that forms after a melted body cools to solid form [36].

• **Silicate Glass**

Any glass whose main component is a silicate.

• **Glaze Level**

Composed of glass, gas, crystal and trace amounts of other raw materials left over after the forming of glaze into a vitreous body. Essentially it is a layer of glass of varying thickness that covers all or a certain portion of the ceramic body [37].

• **Glaze Level Appearance**

Describes the three-dimensional state of the glaze surface; including the topography of the glaze.

• **Glaze Structure Flaw**

Normally refers to split phase areas, unstable areas in the glaze due to the unstable structural integration between two separate phases, or small holes left in the glaze.

• **Macrostructure**

Observable to the naked eye, macrostructure does not

## 相对风化时间

相对风化时间是在绝对风化时间内，在陶瓷整体所处的环境中，陶瓷局部所处的环境发生了改变，导致该处釉层风化作用强度，与陶瓷整体所处的环境中的釉层风化作用强度相比，发生了加速与减速变化的，长短不确定的时间过程。

## 碱金属

碱金属是指在元素周期表中属于第ⅠA族的元素。包括锂（Li）、钠（Na）、钾（K）、铷（Rb）、铯（Cs）、钫（Fr）六种金属元素[49]。

## 碱金属氧化物

碱金属氧化物是碱金属元素和氧元素结合形成的化合物[50]。

## 碱土金属

碱土金属指在元素周期表中属于ⅡA族的元素，包括铍（Be）、镁（Mg）、钙（Ca）、锶（Sr）、钡（Ba）、镭（Ra）六种金属元素[51]。

## 碱土金属氧化物

碱土金属氧化物是碱土金属元素和氧元素结合形成的化合物[52]。

touch on molecules, atoms, electrons, internal-compositions and mechanical structures [38].

• Microscopic Structure

What is observable through the use of an optical microscope, including structures larger than unit cells (the smallest structure of a crystal) [39].

• Microstructure

Any observable structures smaller than a unit cell [40] (i.e. molecule, atom, electron).

• Phase

Refers to the state of an assembled and evenly distributed material that, outside of external forces, has identical physical and chemical properties as well as composition [41].

• Phase Splitting

The process, during the cooling of the glaze level from a melted state, in which materials of certain chemical compositions assemble into split phases [42].

• Vitreous (phase)

Refers to all the non-crystalline composite materials that emerged from a series of physical changes and chemical reactions during firing that solidify when cooled [43]. Also known as glass phase.

• Gas (phase)

The gaseous bodies that are present within the glaze level after the firing of ceramics.

• Crystalline (phase)

Crystallized appearances of certain colors and shapes across the glaze surface or within the glaze level of ceramics. Also known as crystal phase.

• Crystal

A solid body whose internal particles are arranged in cyclical duplication within a three-dimensional space [44].

• Crystal Separation

When a substance separates out from a liquid or vitreous body and forms a distinct crystal phase [45].

• Crystallization

The process of changing from a liquid (whether in solution or melted state) or gas into a solid (under specified physical and chemical conditions) [46].

• Soaking

The phenomenon of a liquid spreading across the surface of a solid when they come into contact; thus reducing the surface energy of the solid [47].

• Weathering

The entire process of physical changes and chemical reactions to the glaze level of ceramics through its matter coming into contact with the surrounding environments [48].

• Weathering Intensity

A measure of the speed and ability of the surrounding environment to break down or change a ceramic's glaze.

### • Weathering Severity

A measure of the weathering strength or weakness experienced by the glaze.

### • Absolute Weathering Time

Refers to the entire passage of time from when a ceramic ware leaves the kiln until it is observed (post-firing time).

### • Variable Weathering Time

Refers to the varied time periods within the absolute weathering time period, in which changes occur to the weathering environment of a ceramic ware; including differences in weathering intensity and environments that can speed up or slow down the weathering process.

### • Alkaline Metal

The elements on the periodic table that belong to the IA family, including: Lithium (Li), Sodium (Na), Potassium (K), Rubidium (Rb), Cesium (Cs) and Francium (Fr) [49].

### • Alkaline Oxide

A chemical compound containing an Alkaline Metal and Oxygen [50].

### • Alkaline Earth-Metal

The elements on the periodic table that belong to the IIA family, including: Beryllium (Be), Magnesium (Mg), Calcium (Ca), Strontium (Sr), Barium (Ba), and Radium (Ra) [51].

### • Alkaline Earth-Metal Oxide

A chemical compound containing an Alkaline Earth-Metal and Oxygen [52].

## 第二节
# 传统陶瓷鉴定依据

## 一、传统陶瓷鉴定依据

在陶瓷迹型学诞生之前，陶瓷鉴定的依据是什么？

李辉柄在《青花瓷器鉴定》中指出，中国瓷器鉴定的年代依据，宋以前的瓷器与明、清以后的瓷器是不一样的[53]：

宋代以前瓷器的年代依据主要是根据墓葬，特别是具有确凿纪年墓葬出土的瓷器，研究者把它作为断代的标准器。考古资料证明，出土瓷器的年代与出土墓葬的年代是相一致的。鉴定家通过对大量由纪年墓出土的瓷器进行科学的排比，找出各个时代瓷器在器型、胎质、釉色、纹饰等方面所具有的不同特征，掌握其发展规律，并以此作为鉴定宋代以前瓷器的科学依据。

明、清时代的瓷器鉴定，因其墓葬很少，特别是带纪年的墓葬更不多见。所以，墓葬不能作为断代的依据。这一时期的瓷器绝大部分都带有年款，如"大明宣德年制"、"大清康熙年制"等，鉴定家把这些带年款的瓷器作为断代的标准器，进行排比研究，然后将带年款瓷器的器型、胎釉特征及纹饰风格作为标尺，用以解决不带年款的官窑与民窑瓷器的断代问题。同时，根据这些带年款瓷器的器型、胎釉及纹饰特征，还可解决一些伪款瓷器的问题。

对宋以前瓷器窑口的断定，必须对其烧造遗址进行发掘，而对其年代的鉴定，要依靠纪年墓出土的遗物，这两者相印证，即可解决墓葬出土物及博物馆收藏品的窑口（即产地）问题。带纪年墓出土瓷器的年代依据，又可用以解决鉴定瓷窑遗址的时代。

以上说明，陶瓷鉴定专家借鉴了考古学中利用类型学原理确定的标准器作为鉴定比对的参照物，所以，目前中国陶瓷鉴定界传统的鉴定依据，可以概括如下：

### （一）利用标准器比对

将考古发掘或传世藏品中已经确定年代及生产窑口的器物作为标准器，鉴定时将之作为参照物，比对被鉴定器物。

### （二）历史文化框架下的思考

在鉴定一件陶瓷器物时，需要陶瓷史知识、陶瓷科学知识、陶瓷工艺美术知识及历史文化知识等，将被鉴定器物放在其被假定的那个历史时期的背景中进行文化思考。因为，处在某个历史时期内的陶瓷器物，它的器型、纹饰、工艺、款识等，均反映了那个时期的政治制度、经济状况、社会习俗、审美理念及艺术追求。这个陶瓷器物就是满载着那个历史时期的信息的物证。反之，那个历史时期的社会风貌和主流文化现象，又是用来判定陶瓷年代的文化依据。如果被鉴定器物身上的文化信息均指向某个历史时期，是属于那个历史时期文化框架内的产物，那么就可以得出被鉴定器物"应该是"那个年代的结论。

所以，专家鉴定器物，只需要将被鉴定器物与已被考古学确定了的标准器进行比对，然后进行历史文化的思考，就能得出结论了。

## 二、关于类型学

类型学是考古界判断器物年代的学问与方法，主要根据古人类所制造的各种器物特征，去识别某一种文化的时代，同时还可以通过分类排比，摸索出它的发展和演变规律，推断年代的早晚[54]。

类型学的奠基人是瑞典人蒙特柳斯，他吸取了《物种起源》中生物学分类进化的原理，将类型学原理运用到考古研究中。他将北欧、南欧的青铜器、陶器与希腊、埃及、西亚的古代装饰纹饰进行比较，对照考古发掘中的一些地层关系，证明有规律可循。

在考古研究领域，类型学已经出现在大学学制教育体系中，但各个学校教材都不统一，都是教师根

# Section 2:
# The Foundations of Traditional Ceramic Appraisal

## 1. The Foundations of Traditional Ceramic Appraisal

Before the emergence of Trace Model Authentication, what was the foundation of ceramic appraisal?

Li Huibing, in his book Underglaze Blue Porcelain Authentication, indicates: The dating foundation for authenticating Song and earlier dynasties' ceramics is different to that of Ming and Qing Dynasty ceramics [53].

For Song and earlier, dating is founded primarily on burial pieces, specifically pieces that are excavated from dated tombs. Archeological reports confirm the dates of the pieces that are excavated from those tombs and burial sites. Appraisers utilize a large amount of dated-tomb excavated ceramics to scientifically classify and find the differences in construction, composition, glaze color, decoration, and other characteristics of different periods; thus grasping the patterns of their development, and then use them as a scientific foundation for authentication of Song and earlier dynasties' ceramics.

Due to the relatively small number of tombs and specifically dated tombs, Ming and Qing Dynasty ceramic authentication cannot be based on the foundation of burial pieces for dating. A large portion of porcelain from these time periods carry the marking of their age: "Made in the Xuande Period of the Ming Dynasty" or "Made in the Kangxi Period of the Qing Dynasty" and so forth. These marked pieces act as the samples experts use in dating and classification research. Then, using the construction, composition, and decoration style of these marked pieces; they carry out dating on unmarked imperial and common kiln pieces. At the same time, they use these marked pieces to face the problem of differentiating forged pieces of the same mark.

Consequently, in determining the kiln of Song Dynasty and earlier ceramics, excavation must be done at the site where the pieces were fired; to determine the date, requires that we use the pieces from dated tombs for reference; both sides are then compared and used to solve the problem of what kiln excavated and museum collection pieces have come from. Ages established on dated tombs can also help to conduct dating on pieces excavated from kiln sites.

The above goes to show that appraisal experts have in fact borrowed the typology concepts from archeology in using a certified sample to act as the consult piece for comparison. Therefore, we can summarize below that the traditional foundation of the current field of ceramic appraisal is:

### A. Using Sample Piece Comparison

The use of archeological excavated or passed down collection pieces with determined dates to act as the samples for consult; thus, using already authenticated pieces for comparison.

### B. Cultural Framework Ideology

When authenticating a piece, it is imperative to use ceramic history knowledge, ceramic scientific knowledge, ceramic industrial art knowledge, historical culture knowledge, and so forth to place the piece in question within the context of the historical period in which it was supposed to have been created. This is because the ware's shape, decoration, technique, markings, etc. are all a direct reflection of the societal convention, economic situation, political landscape, aesthetic desires, and artistic pursuits of that time period. The piece is loaded with messages of testament from that time in history, and the mainstream style and cultural appearances of that period are used as the cultural

据自己的考古实践，总结出理论知识进行传授。不过大学考古系所学的类型学知识，偏重考证陶瓷的生产年代，因为没有考证其真伪的必要。考古发掘与器物真伪的鉴定，是既关联又相互独立的不同领域。

准确地说，考古发掘与陶瓷鉴定不是一回事，考古发掘时利用类型学原理对发掘的遗存器物进行断代，而不需要判断器物真假。

陶瓷鉴定不需要利用类型学原理，只是利用考古发掘的已确认年代的陶瓷作为标准器，将被鉴定陶瓷与之相比对，从而确定被鉴定器物的生产年代是否与标准器相同。被鉴定器物经比对后，如果与标准器相同，可以断定其为真；如果与标准器不相符，可以断定其为新仿或老仿。

## 三、传统鉴定依据的局限性

在当今中国仿古瓷技术已经很发达的历史背景下，传统鉴定的依据已暴露出明显的局限性。当我们用传统的鉴定方法，将出土文物或传世文物作为标准器做鉴定比对时，这些局限性导致了鉴定的准确率下降。致使准确率下降的主要原因有四个：

### （一）鉴定停留在感性阶段

中国陶瓷鉴定自古至今，都是个人的经验性行为，没有统一的客观标准。鉴定知识没有经过学制式的系统学科教育，鉴定水平没有科学统一的衡量标准和考核机构的评定。因此，陶瓷权威鉴定专家李辉柄认为："文物鉴定只是对某件文物进行鉴定，做出年代的判断，然后判断对错与否以及为什么等问题，仅停留在'只能意会，不能言传'的感性阶段上。"[55]

### （二）鉴定停留在人证阶段

传统的鉴定属于感性的经验性鉴定，其鉴定证据是将头脑储存的标准器通过鉴定者口述的模式表达，即人证证据。这种主观表述经验的证据方式，又无法确切证明自己是对的，所以结论只能是或然的。这同当今社会科技高度发达的背景是不相称的。

以上两点均为主观表述为主，缺乏客观标准，因而常有谁也不服谁、谁也说不服谁的情况发生。

### （三）鉴定停留在已有标准器

利用某个时代的标准器或典型器进行比对鉴定，如果有些被鉴定器物的文化信息的指向与标准器发生了冲突，其真实性就会遭到否定。除非这个被鉴定器物有明确的出土年代记载，它才会作为新的标准器纳入数据库，成为新的标准器。

当被鉴定器物遭到否定时，有两种情况：一是确实是现代臆造产品，二是以前没有经国家考古发现的器物，可能属于盗墓或其他来源的器物，但它确实是那个时代的非典型产品。所以，利用标准器比对鉴定，尽管抓住了被鉴定器物应具有的那个时代的普遍性特征，但对于同样属于那个时代的具有特殊性的器物，则缺乏有效认定。

标准器种类不全，会使鉴定结果有局限性，把真品鉴定成非真品的可能性比较大。

### （四）难抵仿古瓷技术攻击

如今仿古陶瓷技术已经能够轻易进入"指定"年代的仿制。

仿真技术发展速度快，主要得益于计算机的参与。计算机技术日益渗透到人类生活的各个层面，正在逐渐改变人类的生活方式。互联网和即时通讯技术的出现，就是一个最好的证明。对于古陶瓷标本上各种肉眼可以观察到的特征，通过计算机的计算和控制，进行仿真，在理论上和实际上都是可以做到的。

传统鉴定注重的比对器物的器型、工艺、胎釉、纹饰、款识等都已被当今的仿古瓷技术击破。现实中已经有不少专家在仿品上"打眼"。

中国目前的陶瓷鉴定依据与证明方式已经暴露出它的局限性。这种局限性导致了陶瓷鉴定的准确率普遍下降。于是，公众质疑专家的鉴定能力，致使中国陶瓷鉴定界的社会公信力受到影响。

那么，如何突破过去陶瓷鉴定依据的局限，提升陶瓷鉴定的公信力呢？陶瓷迹型学的诞生，以创新的思维方法、观察方法及实验方法，以客观物证，促使中国陶瓷鉴定冲破局限，走出困境，重塑公信力！

foundation for determining its age. If the piece being appraised has cultural indicators of a certain time period, they are the product from within the cultural framework of that period. Therefore the conclusion can be made that that piece "probably" is from that time period.

To clarify, when appraisal experts are appraising a piece, they simply need to compare the piece in question with a sample piece of archeologically determined origin; and use cultural framework ideology to form their conclusion.

## 2. Typology

Typology is a systematic branch of learning and methodology used by archeologist to determine the dates of articles. Typology is a method that is specifically used in archeology to primarily categorize all utensils manufactured by humankind; and from that, to understand the culture of that time. At the same time, using classification to feel out the patterns of development, determine the relative age of the utensils [54].

Most archaeologists give Oscar Montelius the credit for the first serious application of the typological method when he transferred the concepts of biological classification found in Darwin's Origins of Species to archeology. He used typological method to categorize the ancient adornments on bronzes and ceramics of North and South Europe as well as Greece, Egypt, and West Asia; and proved that there were patterns in the relationships between the excavation stratums in those places.

In the realm of archeological research, typology is already touched upon in the tertiary educational system, but the curriculum varies from school to school; and each professor draws on his or her own archeology practices to pass on the theories and knowledge. However, the typology knowledge in all tertiary archeology departments stress the textual research of ceramics comparative dating because there is no need to stray into the textual criticism of appraisal. Archeological excavation and appraisal are two, related but opposing, separate domains.

To clarify, archeological excavation and appraisal are two completely different fields. Archeological excavations utilize the concepts of typology to determine the date of excavated articles, and have no need to determine authenticity. Whereas, appraisers do not need to utilize typology, they only need to make use of the already excavated pieces of a certain period to act as a control sample for comparison; looking at the piece in question and determining whether it is of the same production period as that of the control sample. If the piece in question is deemed to be the same as the control piece, it can be judged a genuine; but if it is not the same, it is judged to be a new forgery or a later period replication.

## 3. Limitations in the Foundation of Traditional Appraisal

We hold that at present, with the long history and large developments made in the technology of ceramic forgery, the foundations of traditional appraisal have shown obvious limitations. When we use the traditional methods of using excavated or passed-down pieces for comparison and appraisal, these limitations cause the accuracy to decline. The four primary reasons for the decline in accuracy are:

### A. Appraisal Remaining in a Perception Stage

From the past until the present, Chinese ceramics authentication has relied on an individual's experience; lacking an objective standard. Authentication knowledge had never gone through a systemized set of academic teachings, and authentication ability lacked scientifically measurable standards or a mechanism for examination. As the authoritative appraisal expert Li Huibing puts forth, "Relic authentication is only for dating, and then determining the validity; it doesn't answer the 'why', and it is still remaining in the 'only perceivable by intuition, unable to explain in words' perceptual knowledge stage." [55]

### B. Appraisal Remaining in a Human Witness Stage

Traditional appraisal is merely a perceptual experience model of appraisal utilizing one's memory of a standard piece and being represented through the oral account of human witness. This technique of subjective interpretation of evidence makes it incapable of decisively proving that what is being said is indeed fact. It is inconsistent with the highly developed science and technology backdrop of today.

It is because of the above stated two points, a 'he said, she said' situation arises, in which there is a lack of objective norm.

## C. Appraisal Confined to Already Known Sample Pieces

When compared against a pre-existing typical sample piece, if the cultural indicators of the piece in question are found to be inconsistent, it has to be counted as inauthentic. That is unless the piece itself is excavated and has a clearly known date. In which case, it can be added into the database as a new sample piece for comparison.

A piece that has been deemed a fake has two possible explanations: One, it is indeed a later produced forgery; or two, it was not discovered by the National Archeological Department. It is possible that it is a tomb-raided or other-origin article, and a non-typical case from that period. While the representative sample piece comparison method can catch all the articles in question with the universal characteristics that a certain time period should include; it lacks the ability to effectively capture pieces of the same period that may contain peculiarities.

The lack of a complete sample base limits the authentication results, leading to a higher probability of genuine pieces being deemed as forgeries.

## D. Inability to Withstand the Attacks of Antique Forgers

Today, forgers already possess the technology to easily enter into the forgery of any appointed period.

The high speed at which forgery is now developing is primarily contingent on the incorporation of computer technology. Computer technology has permeated every area of human life, and is gradually transforming our lifestyles; the Internet and instant messaging is probably the best proof of this fact. Now, through the use of technology brought about by computers – both theoretically and practically – all of the naked-eye observable characteristics on ancient ceramics are feasibly forged.

The realm of cultural framework that appraisers use to compare shape, technique, body, glaze, and decoration has already been destroyed by imitation technology. In reality, forgeries already have pulled the wool over the eyes of many modern experts.

Through analysis, we already know that China's ceramic appraisal foundations and certification methods lay bare with many limitations. These limitations lead to the decline in appraisal accuracy. Hence, the public is calling into question the ability of the experts, and the realm of relic authentication's credibility has been gravely affected.

So, how do we break through the foundational limits of the past appraisal methods and regain public credibility in ceramics authentication? The emergence of Ceramics Trace Model Study has already brought forth groundbreaking methods for thought, observation, and testing; while establishing 'seeing-is-believing' objective material evidence. Thus, it breaks through the limitations in China's authentication field to move away from impassable straights and reconstruct public trust.

# 第三节
# 陶瓷迹型学鉴定依据

## 一、陶瓷迹型及规律是鉴定的主要依据

我们对陶瓷到底要鉴定什么？

重点鉴定被鉴定器物的生产年代及生产窑口。同时，也可以对被鉴定器物进行历史价值与文化价值的评述。历史价值可侧重评述该器物对科技工艺的贡献，文化价值可侧重评述该器物的艺术审美价值。

无论是传统的鉴定方法还是陶瓷迹型学的鉴定方法，都要解决这些问题。

其实，传统眼学鉴定也关注到了陶瓷迹型的存在，只是没有更深入地去研究陶瓷迹型问题，而是将陶瓷迹型现象作为鉴定的补充手段。相反，陶瓷迹型学是将陶瓷迹型作为鉴定的主要依据，将传统的鉴定依据作为补充手段。陶瓷迹型学以客观的迹型标准图谱模型比对鉴定，取代传统的标准器比对鉴定。

也就是说，陶瓷迹型学的鉴定依据是：

从思考框架层面上，依据陶瓷在不同存放环境中因风化作用强度快与慢的差异，导致风化强度不一致，在釉表面及釉层中形成的迹型及迹型之间关系的规律特征；

从实际操作层面上，运用迹型标准图谱模型的比对形式进行鉴定，并辅之以逻辑证明、实验证明。

用陶瓷迹型存在的客观实在性和迹型在时间过程中的变化特征及规律作为鉴定依据，可以辨别迹型自然风化的量变特征与人为做旧的突变特征。

## 二、陶瓷迹型标准图谱模型是鉴定的物证系统

在人类司法实践的历史进程中，证明方法曾经有过两次重大转化：

第一次是从以神证为主的证明方法向以人证为主的证明方法转化。

第二次是从以人证为主的证明方法向以物证为主的证明方法的转化。

与此相应，司法证明的历史分为三个阶段：

第一个阶段，神判为证明的主要形式。

第二个阶段，当事人和证人的陈述为证明的主要方法。

第三个阶段，物证或科学证据为证明的主要手段。

现代司法证明就是以物证为主要载体的科学证明。

从神证、人证到物证是一个缓慢的递进过程，是从野蛮走向文明的过程，是人类认知的一个提高过程，最终，物证成为新的"证据之王"。

长期以来，陶瓷鉴定都是采取以专家陈述为证明，你说我听，有神秘的感性色彩，缺乏客观量化、数据化的证据，这样的鉴定结果就会有不确定性。如果说在人类认知水平和科学技术水平很低的阶段，采用这种方法，有特定历史背景的合理性，但是在社会和科技高度发达的今天，仍然采用这种重人证、轻物证的方法，就与时代脱节了。

陶瓷迹型学以陶瓷迹型为鉴定依据，以陶瓷迹型标准图谱模型为鉴定的物证系统，从人证的鉴定时代走向了客观迹型比对的鉴定时代，继而将走向陶瓷迹型自证的时代。

# Section 3:
# The Foundations of Ceramics Trace Model Study

## 1. Ceramic Traces and Ceramic Trace Laws as the Primary Foundation of Authentication

What is it, exactly, that we are looking to authenticate with ceramics?

Primarily we are authenticating the production period and production kiln of the authenticated piece.

Additionally, we can make commentary as to the historical and cultural value of the authenticated piece. From a historical perspective we lay emphasis on a piece's technological impact, while from a cultural perspective we emphasize its beauty and artistic contribution.

Regardless of whether it is traditional or Trace Model Authentication, these questions must be answered.

Actually, traditional intuition-based appraisers have also taken note of the existence of ceramic traces, but have failed to further enter into the research of these ceramic traces. They utilize trace phenomenon as a method of appraisal supplementation. Conversely, Trace Model Authentication utilizes ceramic traces as the primary foundation; utilizing traditional foundations as supplement to authentication. Ceramics Trace Model Study utilizes an objective Standard Trace Atlas Model comparison; replacing the sample piece comparison of traditional appraisal.

This is to say, the foundation of Trace Model Authentication is:

From an abstract framework standpoint, due to the varying weathering intensities of different post-firing environments, and resultant variable weathering severities; it means there are a set of laws that govern the traces and inter-trace relationships across the glaze surface and within the glaze level;

In practice, it is the use of a Standard Trace Atlas Model comparison method of authentication, supplemented with logical and reproducible proofs.

## 2. Standard Trace Atlas Model as the Evidence System for Authentication

Throughout the history of man's legal system, the methodology of proof has undergone two major developments:

First, the progression from primarily Divine Evidence method to primarily Human Evidence method;

Second, the progression from primarily Human Evidence method to primarily Material Evidence method.

Thus, we can divide legal history into three periods:

Period 1: Divination as the primary source of proof;

Period 2: Hearsay or human witness as the primary source of proof;

Period 3: Material or Scientific evidence as the primary source of proof.

Current legal systems adopt a primarily material evidence method for creating proof.

The progression from Divine to Human to Material evidence has been a slow, gradual progress of civilization; finally elevating human knowledge to the point of 'Material Evidence Is King' that we are now enjoying.

For a long time, traditional ceramic appraisal has retained a human-witness based 'I speak, you listen' style of authentication with an air of mystery about it. It lacks objective, quantitative, measurable evidence; leading to ambiguity in ceramic appraisals. If we were still in a period of low knowledge and limited technological ability, utilizing this method would befit the historical background. However, for us to continue to use it, in this socially and technically advanced age, represents a disconnect with our current period in time.

Ceramics Trace Model Study, utilizing ceramic traces as the foundation for authentication and the Standard Trace Atlas Model authentication evidence system, represents the continual progression from a human witness appraisal era to an era of material, evidence-based, objective trace comparative authentication, and even the continuation towards an era marked by ceramic traces acting as self-supporting proof.

## 第四节
# 陶瓷迹型学鉴定原理

## 一、陶瓷迹型现象存在的必然性原理

无论是古代陶瓷，还是现代陶瓷，一出窑，无论时间长短，风化作用都会在陶瓷釉层留下痕迹。

不同的存放时间、不同的存放环境以及存放环境的变化，都会导致陶瓷釉层的风化作用强度和风化强度有差异。

风化作用强度关注的是在某一时间段内，陶瓷釉层被外界环境破坏的速度。

风化强度关注的是陶瓷釉层风化结果的高低程度。

不同的风化作用强度会导致不同的风化强度，这在陶瓷釉层上反映为简单或复杂的迹型特征，它们有着一定的变化规律，体现了时间过程，这就是陶瓷迹型存在的客观必然性、发展必然性及变化的必然性。

正因为陶瓷迹型的客观存在，为我们提供了客观物质观察分析的对象。

## 二、陶瓷迹型量变原理

陶瓷迹型量变原理：陶瓷迹型的风化强度，是在绝对风化时间内，以缓慢、渐变的形式形成的，其中包含无数个在相对风化时间内形成的风化强度。

陶瓷釉层有着极强的抗风化能力，因此，陶瓷的风化过程是缓慢地进行的，这与当今陶瓷仿制技术中的人为做旧有着本质区别。前者是量变特征，后者是突变特征。尽管现在高科技做旧，能够做出某类化学分子结构相同的迹型，但人为做旧造成的突变特征与陶瓷釉层自然风化的物理堆积过程和化学溶蚀后的形貌特征，还是截然不同的。

陶瓷迹型的量变原理，在陶瓷的绝对风化时间与相对风化时间的过程中都有反映。只有在绝对风化时间过程中去认识领会迹型的量变特征，在相对风化时间过程中认识领会风化作用强度不一致现象，才能理解迹型最终风化强度的多样性和复杂性。

在绝对风化时间内，迹型从无到有，从不明显到明显，包含了在许多相对时间内风化作用强度不一致的现象发生，其中渗透着质量互变过程。

正因为如此，我们才能观察到具有风化的绝对性、相对性、复杂性的特征，风化强度有差异的迹型。这就是迹型发展、变化的辩证法，这就是我们观察分析迹型现象的核心指导思想。

## 三、陶瓷迹型变化时间过程原理

陶瓷迹型变化的时间过程原理是：在绝对风化时间内，陶瓷迹型有着生成、发展的时间过程。在这一过程中，穿插伴随着许多相对风化时间，它们使陶瓷釉层各处的风化作用强度不一致，最终形成了陶瓷迹型风化强度的多样性和复杂性，但量变特征贯穿始终。

要深刻理解这一原理，必须要理解陶瓷的绝对风化时间与相对风化时间。

如何在绝对风化时间内认识迹型变化的时间过程性？在外界环境稳定均匀的前提下，风化强度的高低取决于釉层各部位的成分差异以及结构稳定性，成分的物理化学性质较活跃、结构稳定性差的釉层先风化出我们可观察到的迹型，因而风化强度高；而成分的物理化学性质较稳定、结构稳定性好的釉层后风化出我们可观察到的迹型，因而风化强度低。迹型风化强度的差异就体现了时间过程。

风化强度的差异可以反映釉层迹型的生成、发展的时间先后轨迹。所以在绝对风化时间内，如果陶瓷整体所处的环境稳定均匀，就可以得出这样的结论：陶瓷某一迹型的风化强度越高，说明迹型形成的时间越长，这是陶瓷迹型存在的

# Section 4:
# The Principles of Ceramics Trace Model Study

## 1. The Inevitability Principle

Regardless of ancient ceramics or modern productions, once a piece has left the kiln, no matter for how long, weathering has begun to leave traces upon the glaze of the ware.

Differences in passage of time and surrounding environments as well as changes in those environments all bring forth variations in the weathering intensity and weathering severity.

Weathering intensity refers primarily to the speed at which a piece is deteriorated within its external environment over a period of time.

Weathering severity looks at the high or low level of weathering outcome on the glaze level.

Different levels of weathering intensity directly transpose into differences in weathering severity; reflected in a range of simple to very complex trace characteristics within the ceramic glaze level. Their changes, occurring within specific laws and patterns, reflect a passage of time. This is the objective inevitability, development inevitability, and change inevitability of ceramic traces.

The existence of ceramic traces gives us an objective target material for observation and analysis.

## 2. The Quantitative Change Principle

The weathering severity of ceramic traces occurs within an absolute weathering time. It is a slow, gradual formation that is the accumulation of weathering severities formed during innumerable variable weathering time periods.

Ceramic glaze has an extremely high resistance to weathering forces. Because of this, the traces left after hundreds of years of trace weathering processes, progressing very slowly, are innately different to those brought on in a sudden aging process produced by forgers. The first having quantitative change characteristics, while the latter has sudden change characteristics. Even though some modern forging techniques are able to produce identical chemical compositions; the man-made forging sudden change characteristics and ceramic glaze level natural weathering quantitative physical buildup and chemical corrosion process characteristics are entirely different.

The principle of ceramic trace quantitative change is reflected in both the absolute weathering time as well as the variable weathering time periods. It is only through recognizing the quantitative change characteristics of the absolute weathering time and the irregularity of the weathering intensity during variable weathering time periods, we are able to understand the multiplicity and complexity of the final trace weathering severity.

Within the absolute weathering time, traces form where there were none before, and gradually gain distinctiveness. It includes many variable time periods with inconsistent weathering intensities; entering into a process of mutual change of essence and quantity.

It is specifically because of this that we are able to observe the absolute, variable, and complex differences in the trace weathering intensity. This allows for a dialectal logic method for looking at trace transformation and changes, and is the core guiding principle in the observation and analysis of ceramic traces.

## 3. The Passage of Time Principle

The passage of time principle is: Throughout the absolute weathering time, the ceramic trace generation and transformation processes, within numerous variable weathering time periods,

普遍现象和特征。而陶瓷存放环境变化越少，这种结论就越可靠。

但这在现实中是少见的，陶瓷存放的环境，无论在水中、地下，或空气中，或多或少会发生变化。当陶瓷存放环境发生了改变，例如一件存放于地下的陶瓷，地下发生了塌方或渗水，陶瓷一部分浸泡在水中，那么有水浸泡的釉层，其风化作用强度就会与无水浸泡的釉层有差异。

由于陶瓷局部的存放环境改变，从改变的那一刻起，改变的环境所对应的釉层就会发生风化作用强度的变化，这就是迹型相对风化时间的开始。可以说，在现实环境中，在迹型的绝对风化时间内，都会包含多个相对风化时间，而且，一件陶瓷上的不同部位可以经历不同次数的环境变化，也就是经历的相对时间数量不同。比如：一件陶瓷的肩部所处环境变化次数较少，其经历的相对时间数量较少，而圈足所处的环境则不断在变化，经历的相对时间数量就比较多。这些相对时间是无法确定起始和结束的时间段。

所以，针对陶瓷釉层整体的迹型而言，不能笼统地断定风化强度高的迹型形成时间一定长，风化强度低的迹型形成时间一定短。因为这涉及到相对时间的出现问题。对于同一件陶瓷上的迹型而言，如出现釉层某一局部的风化强度总体高于或低于其他部位的现象，这仅仅表明，该局部与其余部位的风化作用强度有差异，并不一定代表该处所有迹型形成的时间比其他部位的迹型更长或更短。

那么，如何在相对风化时间内认识迹型变化的时间过程性？相对时间的开始，意味着陶瓷局部所处的环境发生了变化。就这一局部而言，所有迹型的风化强度的差异，可以由以下原因造成：

一是该釉层局部各处成分的物理化学性质有差异。

二是该釉层局部各处结构稳定性有差异。

也就是说，成分的物理化学性质较活跃、结构稳定性差的釉层先风化出我们可观察到的迹型，风化强度高；而成分的物理化学性质较稳定、结构稳定性好的釉层后风化出我们可观察到的迹型，风化强度低。迹型风化强度的差异也能够体现时间过程。

我们目前尚不需要证明各种迹型出现的先后顺序，只需要观察各种迹型最终的风化强度，因为对陶瓷迹型年代的判断依据就是：陶瓷各种类迹型在绝对风化时间过程中形成的风化强度的普遍性特征。

此外，各种迹型的多样性与复杂性，主要是相对风化时间留下的痕迹。陶瓷存世的时间越长，意味着陶瓷存世环境的变化可能性越高，也就意味着迹型的相对风化时间越多。通过观察迹型在相对风化时间过程中的特征，可以进一步了解器物存放环境变化的复杂性。

所以，时间过程原理告诉我们，每一个迹型单项的最终风化强度，都经历了同样的绝对时间过程，而相对时间的介入，无非是使迹型风化的速度加快或放慢了而已，无非是更准确地描述了迹型发展变化的复杂现象，讨论相对时间的意义仅仅在于：它能够帮助我们理解迹型变化的快与慢以及迹型的多样性、复杂性。而我们要关注的核心问题是：迹型的多样性与复杂性所表现出的量变特征。

现代高科技做旧陶瓷，是通过人为刻意改变陶瓷的生存环境，在很短的时间内，造出风化强度很高的迹型。但这种方法难以造出迹型在多个相对风化时间内形成的多样、复杂的量变特征。

form the multiplicity and complexity of trace weathering severities throughout the ceramic glaze level. However, quantitative change characteristics are universally present throughout.

To deeply understand this principle, we must understand the concept of absolute and variable weathering time.

How do we recognize the passage of time in trace changes within the context of absolute time? In a stable and constant environment, weathering severity is determined by the structural stability of the glaze, as well as the chemical composition in each spot. Areas of low structural stability and highly active chemical compositions are first to present with observable trace changes, thus have a higher weathering severity; and portions of high structural stability and relatively inactive chemical compositions are the last to present with observable trace changes, thus have a lower weathering severity. These differences in weathering severity represent a passage of time.

Variations in weathering severity also reflect the passage of time in glaze level trace change, and are product of the trace generation and transformation process. Therefore, if within the absolute weathering time the post-firing environment is stable and constant, we can put forth the following assertion: The higher weathering severity of a certain ceramic trace, the longer the passage of time in that environment; this is the common phenomenon and characteristic of all ceramic traces. We can further say that the lesser the amount of changes in the post-firing environment, the truer the previous assertion holds.

However, the reality is that this constant is rather rare. The majority of post-firing environments, regardless of whether it is underwater, underground, or in the open air; will undergo change. When the environment changes, for example underground wares experiencing landslide or flooding and a single piece becomes partially soaked in water, then the portion of glaze level that is soaked in water will have a different intensity of weathering than that of the portion not soaked.

Due to the changes in environment to a ceramic ware, from the moment of that change, the changed environment impacts on that piece a change in the weathering intensity, thus beginning a new period of variable weathering time.

This is to say, due to the nature of changing environments, any one absolute weathering time period contains several separate variable weathering time periods. Also, any one piece can present with different amounts of environmental changes, which is to say different amounts of variable time periods, at each point on that piece; i.e the environment of the shoulder could have remained relatively constant, whereas the environment of the foot ring may have changed repeatedly. Additionally, due to the changes of weathering intensity across variable weathering time periods, we are unable to assert the length of time of trace generation and transformation in any one specific variable weathering time period.

Further, speaking of all the traces within an entire glaze level, we cannot simply assume that higher weathering severity reflects a longer passage of time and lower weathering severity is indicative of a shorter trace formation period. This comes down specifically to the issue of the emergence of variable time periods. If a spot on the porcelain appears to have traces with relatively severe or less severe weathering, it is because the weathering intensity is higher or lower in that location on the piece, and does not necessarily mean that all the traces within that spot are formed over a longer or shorter period of time than other points on the piece.

Now, how do we recognize the passage of time in trace changes within the context of variable time? The start of a variable time period is marked by a change in the post-firing environment at any specific point. The differences in weathering severity at that point are determined by the following:

Differences in the active nature of the chemical composition; or

Differences in structural stability.

Areas of low structural stability and highly active chemical compositions are first to present with observable trace changes, thus have a higher weathering severity; and portions of high structural stability and relatively inactive chemical compositions are the last to present with observable trace changes, thus have a lower weathering severity. These differences in weathering severity also represent a passage of time.

Now, it is not necessary to determine the order in which

traces appear, but merely observe the final weathering severity of each trace. Therefore, the basis in determining the age of a ceramic ware is: All the ceramic trace types formed during the absolute weathering time; which is its representative trace characteristics.

Additionally, the multiplicities and complexities of all traces are primarily the byproduct of variable weathering time. The longer a ceramic ware has been in existence, the higher the probability of more environmental changes, which would translate as more variable weathering time periods. Through observing the trace appearances brought on by variable weathering time, we can further understand the complexity of the post-firing environment.

The passage of time principle essentially tells us that the weathering severity of all individual traces is brought about during the same absolute time. The involvement of variable weathering time periods helps us understand how these variations sped up or slowed down the weathering process and brought on multiplicity and complexity in the development and change of traces. What we really need to focus on is the quantitative change characteristics of this multiplicity and complexity.

Modern high-tech forgeries can be said to have undergone a deliberate change in the post-firing environment. Within a very short period of time, very high weathering severity is generated. However, these techniques cannot produce the complex and multiple types of traces formed under the varying weathering intensities brought upon by variable weathering time which characterizes quantitative change.

# 第五节 陶瓷迹型学鉴定本质

陶瓷迹型鉴定本质，就是陶瓷迹型鉴定方法自身的属性特点，与其他鉴定方法的不同之处。

## 一、观察整体化

观察整体化就是要求在鉴定器物时，要对被鉴定器物可能被观察到的每一个部位都要观察，从而将观察到的所有迹型特征记录下来，以备量化分析。

科学的鉴定，就是建立在量化基础之上的。

在鉴定实践中，有时仿品上的迹型与真品难以区分，这就需要观察器物的整体，将器物所有的迹型特征记录、归纳、分析，从而得出结论，不能仅仅凭局部几张图片来判断真假。

传统眼学鉴定的思路是求异化的，即被鉴定器物在比对标准器时，若发现有一点不符合标准器上的特征，被鉴定器物就不被看好了。但是，不看好的器物，有两种情况：其一，是仿制得不到位，确实是假；其二，是由于在传统眼学鉴定中人知识的有限性，导致鉴定者将认识不清的信息误认为是不符合标准的信息。

传统鉴定经常可以听到这样的说法："其他信息都达标，只有一点不达标。"然后就根据这一点不达标的信息，否定该器物。其实要否定该器物，不能这么说，因为真品上的信息，要么都真，仿品上的信息要么都假，不可能存在只有一点假，其他都真的现象。

关于这一点，无论是传统眼学鉴定与陶瓷迹型鉴定都应该是一致的。

在利用陶瓷迹型学原理进行鉴定时，前提是观察整体化。尽管只要证明一种迹型是真实可靠的，就可以推及其他迹型都是真实可靠的。但是，去发现更多的迹型证据存在，去揭示更多的迹型存在规律以及迹型之间的关系规律，并证明它们之间逻辑关系的客观实在性，这样，更多的靠谱迹型，根据彼此之间的必然联系，相互支撑，使得比对鉴定更有说服力。

比对标准图谱模型的思路是求同化、客观化的。如果发现被鉴定器物有多种迹型与标准图谱模型有本质性相似，高度靠谱，而且这些迹型种类目前还难以仿出，那么，这件器物就可断真。

只有整体化去观察，才有可能找到最接近标准图谱模型的更多迹型，才能不被局部的迹型假象所迷惑。

## 二、比对证据客观化

陶瓷迹型标准图谱模型，不是鉴定者头脑中储存的标准器表象的经验性记忆，而是在现实中发掘、提炼的客观存在现象。陶瓷迹型标准图谱模型是客观证据的典型化、集中化、系统化。

模型不仅在科学研究中使用，在现实社会中也广泛存在。比如，人类亲子鉴定的基因图谱模型，司法鉴定中的笔迹鉴定，武器中的瞄准系统，地图也是寻找方位的客观模型。

在观察大量的有明确发掘时间、发掘地点及年代依据或有明确传世记载的陶瓷迹型后，建构出简练、概括的陶瓷迹型标准图谱模型。它是客观的、毋庸置疑的物证系统，借助它的客观性，并结合逻辑推演证明，从事陶瓷的比对鉴定。这种方法属于客观的物证范畴。

不同的陶瓷标本迹型能够提炼出不同的迹型图谱模型，因为它们分别反映了釉料不同的配方和不同的工艺。这时，就需要将迹型特征本质相同的迹型分类，建立不同"家族"的迹型标准图谱模型，使得不同用料、不同工艺的陶瓷均能得到有效鉴定。

在研究的初级阶段，陶瓷迹型学鉴定是利用陶瓷釉层的客观迹型特征及它们之间关系的规律为鉴定依据，运用迹型标准图谱模型比对被鉴定器物，然后做出推理与判断。

# Section 5:
# The Nature of Ceramics Trace Model Study

The nature of Ceramics Trace Model Study lies in the innate characteristics of its methodology, and is what sets it apart from other authentication methods.

## 1. Holistic Observation

Holistic Observation refers to the necessity for every observable part of the piece in question to be observed during authentication. Through this form of observation, all trace characteristics are recorded for quantitative analysis.

Scientific authentication is firmly footed on a quantitative foundation.

When carrying out authentication, it is sometimes difficult to differentiate the traces of real and fake pieces. Forgeries can present with trace characteristics similar to real pieces; which requires our observation to be holistic. We must survey an entire piece and record all trace characteristics, categorize them, and analytically put forth a conclusion; not merely lean on a few images for final differentiation.

During traditional intuition based appraisal, a disconnecting framework is required. If any characteristics in the piece being appraised are not congruent with the sample counterpart, it is not appraised favorably. However, there are two possibilities to these unfavorable pieces: One, is they are incorrectly forged, and in fact a fake; or two, it is due to the limitations in the intuition of the appraiser that a point not recognized is mistakenly seen as an inconsistency, and the piece is found unfavorable.

It is common in traditional appraisal to hear "It has some consistent indications, but it is just slightly inconsistent" and because of this irregularity; the piece is deemed a forgery. In fact, when deeming a piece fake, we should not utilize this logic; either the piece is entirely real or it is entirely fake, we cannot say it has a subtly fake indication, but is mostly real. This fact should hold true regardless of whether we are using intuition based appraisal or Trace Model Authentication.

When using Ceramics Trace Model Study principles for authentication, as mentioned before it is imperative that we are holistic. Even though we only need to find one trace characteristic that is reliable, we can put forth that the other traces are also reliable. However, discovering more trace evidence allows us to further put forth even more trace and inter-trace relationship laws; as well as logically prove the concrete reality of these relationships. This way, we have more dependable traces that are positively matched and mutually supportive, bringing about increased persuasiveness in comparative authentication.

Standard Trace Atlas Model comparison requires an assimilative and objective framework. If a certain trace type is found to be consistent with that of the atlas, and this trace type is currently infeasibly faked, this piece can be confirmed as real.

Only through holistic observation, the most atlas-consistent traces can be found, thus avoiding confusion by any traces with fake appearance.

## 2. Objective Evidence

The Standard Trace Atlas Model is not the left over memories in the appraisers mind from past experiences of sample pieces. Rather, it is extracted and refined from objective fact. The Standard Trace Atlas Model is the typification, centralization, and systemization of objective material evidence.

In fact, models are not merely used here in scientific research, but are present throughout many aspects of our life.

迹型标准图谱模型弥补了在陶瓷传统鉴定时，只关注器物形制、工艺及文化信息的普遍性，而对它们的特殊性欠缺考虑的缺陷。陶瓷迹型学的鉴定研究，关注的不只是肉眼所能观察到的器物信息，更重要的是关注仪器才能观察到的显微迹型信息。而赝品在显微镜下呈现的迹型在图谱模型面前无所遁形，因为这些迹型标准图谱模型展示出来的迹型信息，就是无法否定的客观物证。

传统眼学鉴定以人为本，以人的观点述说为证据，而陶瓷迹型学鉴定则以迹型标准图谱模型为本，通过迹型标准图谱模型比对，以迹型本质的相似度进行推理判断，以客观的迹型为证据，以迹型之间的关系规律为证据，使鉴定的主观性变为客观性，鉴定的人证变为客观比对的物证，人的主观经验性鉴定行为降为次要地位，这使得鉴定判断从或然性走向实然性。

客观性迹型标准图谱模型，属于自然科学中的客观现象驱动模型系列，不像社会科学中的许多理论驱动模型建构。如：在经济计量学中，假定先验理论模型的正确性，将计量模型仅作为测度理论模型参数估值的工具，这就有可能出现先验的理论模型对现实中的经济动荡与危机的预测失误。因此才会有更客观的动态经济计量学诞生，这样就能更加准确地预测未来的经济现象。

陶瓷迹型标准图谱模型，是建立在有明确发掘时间、发掘地点及年代依据或有明确传世记载的陶瓷标本之上的，因此，它的证据客观性是毋庸置疑的。

## 三、论证逻辑化

鉴定与鉴别、鉴赏、品鉴不同，鉴定不仅仅是用嘴说话，更要用笔说话，要有凭据，不能仅仅靠说理服人，要立字为据，鉴定论证时要有充足理由的推理原则。要达到理性认识阶段的鉴定认识，就要运用概念、逻辑去推理判断。

这些思维内容，在客观证据真实的条件下，可间接概括性地反映事物特征，还能够以这种真实的直接知识为中介，推出新知识，提出科学假说，以求实验证实。整个鉴定过程，要求符合思维的逻辑规律，达到有条理、易理解的效果。

## 四、求证实验化

陶瓷迹型鉴定，是以陶瓷迹型风化的量变特征，以及迹型之间变化呈现的时间过程性为基本原理的，那么就可以通过试验，去模仿陶瓷迹型风化的量变特征，以及迹型之间变化所呈现的时间过程性。但是，这种试验永远也不可能得到与数百年量变的迹型风化特征一致的结果，只能无限"像"地去接近，这种"是"与"像"的差异特征就是求证判断迹型量变真实性的客观证据。这种实验方法，可由第三者去检视，如此才能进行统计学上的可信度分析。

For example: Human Paternity Test Gene Mapping Models, Expert Witness Handwriting Appraisal Models, Weapons Sighting System Models, and even Maps, which are simply models for finding locations.

Through the observation of traces on numerous ceramic pieces with known excavation record and age or of concrete provenance; a simplified, representative Standard Trace Atlas Model is constructed. It is an objective, reliable material evidence system. Utilizing its objectivity, along with logically deduced proof, we carry out ceramic authentication; a technique categorized within the objective material evidence model system.

Different ceramic sample traces produce different atlas models, this due to their dissimilarities in glaze composition and crafting techniques. Thus, classification according to the inherently different glaze level characteristics must be done in order to create 'family' specific atlas models. This way, ceramics of all glaze compositions and crafting techniques can effectively be authenticated.

The entry level stage of Ceramics Trace Model Authentication is based upon the foundation of the objective laws that govern ceramic traces and their relationships with one another. It first utilizes a Standard Trace Atlas Model for authentication comparison, and then puts forth deductions and findings.

When using traditional appraisal methods in supplement to the Standard Trace Atlas Model, we need to primarily consider the consistent shape, craftsmanship, and cultural indicators; while keeping in mind peculiarities or flaws. Ceramics Trace Model Study and Authentication not only looks at these indicators, observable by the naked eye, but also stresses the microscopic indicators only observable through use of special observation instruments. Among these indicators, many are infeasible to fake. These microscopic indicators in the Standard Trace Atlas Model act as verifiable, objective material evidence.

Traditional appraisal is rooted in the individual and their ability to express their views as proof; but Trace Model Authentication is rooted in an atlas model and its objective material evidence for the comparison and discernment of the validity of each piece. It utilizes the laws of trace relationships, transfers the authority of authentication to objectivity, transforms from human witness to material evidence (placing human witness as supplemental), and moves from appraising possibility to authenticating reality.

The objective Standard Trace Atlas Model belongs to the natural science objective phenomenon-driven model series; dissimilar to that of social science multiple theory-driven abstract model. For example, the priori of Econometrics assumes: The accuracy of a theoretical model is merely used as a parameter tool for estimation in theoretical measurements. This in turn can lead to the misforecast of the theoretical model seen in economic turmoil or crisis. This causes the emergence of more objectivity in Econometrics, bringing about a higher degree of accuracy in future economic phenomenon.

The Standard Trace Atlas Model is constructed on samples of clearly known excavation record and age or of concrete provenance; this way, its objective material evidence is unfaltering.

## 3. Logical Arguments

Authenticating, differentiating, appreciating, and appraising are all different. Authentication goes beyond the dull grey of 'he said, she said' and enters into the distinct black and white. It needs proof more than argument, requires contractual certification, and is proved in accordance with clear-cut reasoning principles. For rational thought to exist, authentication knowledge must be based on defined concepts and logic for rendering judgment.

This type of thought, under the constraints of objective evidence, can reflect the evidential generalities and use the direct recognition of these truths to introduce new knowledge and put forth scientific hypotheses to test their veracity. The entire authentication process demands presentation in accordance with thought in order to reach orderliness and understandability.

## 4. Experimental Verification

Ceramics Trace Model Authentication is based upon the foundational principles of the quantitative change characteristics of trace weathering and inter-trace relationships, incurred throughout the course of absolute weathering time. We can, through experimentation, attempt to fake these quantitative change characteristics of trace weathering and the passage of time. However, these experiments have yet to attain the quantitative change gained only over time, but merely 'resemble' these changes. This disparity between 'is' and 'resembles' is precisely the objective evidence used in the judging of quantitative trace change legitimacy. This type of testing method must now, however, undergo further third-party participation, thus allowing for statistically reliable analysis.

# 第六节
# 陶瓷迹型学局限性

## 一、标本选择的局限性

与眼学中使用标准器存在局限性一样，迹型学中使用标本也存在局限性。

特定时期的陶瓷迹型标准图谱模型，既有无可怀疑的客观性，又有高度的特定性，因而必然具有被新标本证明缺陷的可能性。也就是说，特定时期的陶瓷迹型标准图谱模型，不能识别所有真器，对部分真器可能只能给出待研究的结论。

例如元代青花瓷器的迹型标准图谱模型，是从元代景德镇地区生产的瓷器标本迹型中归纳总结出来的。不同用料及不同工艺的陶瓷具有不同的迹型特征。因此，这个标准图谱模型就不能作为元代其他窑口陶瓷的比对参照标准，不具有通用性。

所以，目前的陶瓷迹型标准图谱模型未来面临被批判的挑战。但是构成挑战的批判正是科学研究的生命线，它意味着陶瓷迹型标准图谱模型有不断发展与完善的可能。

## 二、显微研究的局限性

目前我们观察与研究陶瓷使用的是500X数码显微镜，然而即使是500X显微镜下的成像，有些迹型还是观察得不够清楚。如果能使用更高倍数的显微镜，迹型现象研究将会更精确、更有成效。

如果有更先进的取样和拍摄仪器，可以更加精确分析迹型的成分和结构，鉴定结果也将更加准确。那样，陶瓷迹型学将从目前初级阶段的比对求证走向高级阶段的自证。

另外，由于缺乏更多的整器作标本，整器迹型全方位扫描的数据库的建立是不完善的。还有待在未来不断丰富量化数据库。

人们观察陶瓷釉层现象的测度不是一成不变的，观察的手段与工具也不是一成不变的，这说明了科学的研究与发展是一个充满曲折、螺旋上升的过程。

传统眼学与陶瓷迹型学，是不同的时代的产物，正如工业时代催生了管弦音乐，电子信息时代催生了流行音乐。

从传统眼学研究发展到陶瓷迹型学研究，是社会发展与人类认识发展的必然现象，是人类认识的逻辑递进。人类借助科学仪器，观察器官得到延伸，人类的认识进入了更广阔的视野，在从必然王国走向自由王国的途径中获得了更多的自由。从宏观研究深入到显微研究，再从显微研究深入到微观研究，这就是不断地突破陶瓷鉴定研究的局限性。在不断突破的过程中，从建模的实践中，不断发现迹型标准图谱模型的系统性规律，使陶瓷鉴定向更加量化、标准化、科学化的方向发展，这就是陶瓷研究与鉴定未来发展的方向之一。

依据陶瓷迹型学原理去鉴定，得出的评估与鉴定只是在目前条件下的相对认识。正是这些相对认识所表述出的相对真理，最终才能汇成绝对真理的长河。

## 三、人工操作的局限性

尽管迹型标准图谱模型是客观的，风化留下的迹型就是物证，但在鉴定比对时，还是要有人来操作。这就会出现鉴定结论上的差异，从而影响对被鉴定器物判断的准确性。正如前言所说，我们不缺乏风化迹型的物证，缺乏的是发现这些物证的眼睛。

我们不能因为打不中十环，而去否定瞄准系统；不能因为我们对笔迹鉴定方法未能完全掌握，而去否定笔迹鉴定。

所以，利用陶瓷迹型标准图谱模型比对鉴定，有一个艰苦的学习过程，有一个对客观真品迹型的认识掌握过程。只有对真品的风化迹型认识理解越清楚，掌握越全面，才能在比对鉴定中，"瞄得更准"，获得更多的认识自由，提升

# Section 6:
# Limitations of Ceramics Trace Model Study

## 1. Limitations of Sample Selection

Similar to the limits of sample selection in intuition based study, Ceramics Trace Model Study also faces sample selection limitations.

The Standard Trace Atlas Model for a specific time period has undoubtable objectivity with a high level of specificity, but is illuminated with the possibility of errors shown by future discoveries, thus putting it up for scrutiny and criticism. This is to say, an atlas model of any given period cannot possibly encompass every genuine piece in that period. Some pieces may receive a verdict of 'awaiting further research'.

For example: The Yuan Underglaze Blue Porcelain Standard Trace Atlas Model is based on samples of Yuan Dynasty Jingdezhen Kiln Area Underglaze Blue Porcelain. Because of this, these atlas models are insufficient in the comparative analysis of other kiln sites of the same period. Therefore, the YUB Porcelain Standard Trace Atlas Model cannot effectively be used in the determining of age and validity of ceramics from other areas, i.e. Longquan Kiln. Thus, it is inherently limited.

Therefore, the current Standard Trace Atlas Model will face future challenges. The presence of such critiques is the lifeblood of scientific research. It implies the possibility for the continual improvement and perfection of any atlas model.

## 2. Limitations of Microscopic Research

Currently, we are utilizing a 500X magnification digital microscope for the observation and research of ceramics. However, of the appearances observed under 500X magnification, some can still be unclear. If we were able to attain a higher magnification scope, the clarity of trace appearances would improve, bringing even more fruitful research.

With the emergence of more advanced apparatuses, we can clearly analyze the composition and structure of traces, and authentication can move in a more precise direction. In which case, Ceramics Trace Model Study can move from an entry-level of comparative evidence into an advanced level of self-supporting proof.

Additionally, due to the lack of complete-ware samples, the complete-ware scan database is deficient. A continual need for a more robust sample area and database exists.

The amount to which we observe the ceramic glaze level is not once-off. Likewise, the techniques and tools used for observation are not one-size-fits-all. This goes to show the complicated nature of scientific research and development; a continually upward spiraling process.

Traditional appraisal and Trace Model Authentication are the products of two different eras; just as the industrial age brought about orchestral music, and the information age brought about electronic music.

The progression from traditional appraisal to Trace Model Authentication is a necessity in the progress of society and human knowledge. It is a logical step towards deeper understanding. Mankind draws support from scientific instruments; observation apparatuses receive extension, and our field of vision broadens; putting us on the way from the realm of necessity to the realm of freedom thus gaining more freedom. By moving from macroscopic observation and deepening into microscopic, and then microscopic observation deepening into microcosmic research, we are continuously breaking through the limitations of research. During the process of continuous breakthroughs and carrying out model construction, we continuously discover the systemized

鉴定的准确率。

## 四、陶瓷迹型缺乏的局限性

如果被鉴定陶瓷器物具有体积较小、年代较近（50年以内）、瓷胎较薄等特征，通常其迹型风化痕迹不明显且迹型种类有限，这都会影响陶瓷显微迹型的鉴定。

laws that present within the Standard Trace Atlas Model. We also take ceramic authentication through a quantitative, standardized, and scientific progression; this is the future direction of Ceramics Trace Model Study and Authentication.

Authentication based on the principles of Ceramics Trace Model Study only offers assessment according to the relative knowledge of current conditions. Only from the relative truths explained by this relative knowledge, can we harness an endless flow of absolute truth.

## 3. Limitations of Manual Operation

Even though the Standard Trace Atlas Model is objective, and even though the traces left by weathering are material evidence; comparative authentication still needs to be carried out through human efforts. This will produce inconsistencies in authentication conclusions, and can influence the accuracy of material evidence findings. Just as it mentions in the Foreword of this book: "Ceramics Trace Model Study has never lacked the objective material evidence that ceramic traces offer, it has only lacked the collective insights to discover them."

We must not blame the gun's scope for missing the bulls-eye; or in light of not grasping the handwriting recognition method, negate handwriting appraisal.

In the same way, utilizing the Standard Trace Atlas Model for authentication is an arduous studying process; a progression towards grasping an understanding of objective, genuine traces. It is only when weathering traces are understood more clearly and grasped more completely that our authentication can "hone its sights"; increasing our comprehension and elevating our authentication precision.

## 4. Limitations of Lacking Traces

For pieces with a relatively small surface area, relatively young age (within 50 years), or relatively thin body; it is common for the trace weathering characteristics to be indistinct or to present with a limit in trace types. These factors all affect the Trace Model Authentication of the ceramic ware.

第四章

# 陶瓷釉层的化学成分、性质与形貌

**阅读提示：**

　　釉是一种材料。材料是人类用于制造器件的物质。人类的发展史，可说是人类创造材料、利用材料的历史。釉的发明是陶瓷发展史上的里程碑。釉的发明推动了人类社会的进步和生活质量的提高，就其对人类社会发展的影响力来说，是人类对材料进行创造与利用的一次革命。

　　陶瓷不同的釉料配方、不同的烧制工艺，会使陶瓷釉层表现出不同的迹型特征。这些迹型特征，是我们观察与研究的客观对象。本章将对陶瓷釉的化学成分、性质以及釉层形貌做理论性论述与阐明。

**主要论述问题：**

1. 陶瓷釉层的化学成分有哪些？它们各自的作用是什么？
2. 陶瓷釉层为什么会风化？
3. 陶瓷釉层的形貌是怎样的？

### Chapter 4

# Chemical Composition, Characteristics and Appearance of the Ceramic Glaze Level

**Text Note:**

Glaze is a type of physical material; one which humankind has used in the construction of vessels. The history of human development can be seen through our ability to create and use materials. Glaze is one of the fundamental milestones in the development of ceramics. The invention of glaze pushed forward the progress of our society and elevated our quality of living. It revolutionized our ability to manipulate and utilize materials.

The different ceramic glaze compositions and firing techinques bring forth different trace characteristics; these trace characteristics are the focus of our objective observation and research. The purpose of this chapter is to expound upon and explain the chemical composition, characteristics, and appearance of the glaze level.

# 第一节
# 陶瓷釉化学成分

## 一、釉的概念

釉是施于陶瓷坯体表面的利用天然矿物原料按比例配合，在高温作用下熔融而覆盖在坯体表面的玻璃质层[56]。

## 二、釉的构成

釉主要由网络形成剂、助熔剂和着色剂组成。

### （一）网络形成剂

玻璃相是釉层的主相，形成玻璃相的主要氧化物是二氧化硅（$SiO_2$），它在釉层中以四面体的形式相互结合为不规则网络，所以$SiO_2$又称为网络形成剂[57]。

### （二）助熔剂

在釉料熔化过程中，这类成分能促进高温分解反应，加速高熔点晶体（如$SiO_2$）化学键的断裂和生成低共熔物。助熔剂还起着调整釉层物理化学性质（如力学性质、膨胀系数、粘度、化学稳定性等）的作用[58]。

常用的助熔剂化合物有氧化钠（$Na_2O$）、氧化钾（$K_2O$）、氧化钙（$CaO$）、氧化镁（$MgO$）、氧化铅（$PbO$）等。

### （三）着色剂

着色剂是在陶瓷的胎、釉中起呈色作用的物质[59]。着色剂包括氧化亚铁（$FeO$）、三氧化二铁（$Fe_2O_3$）、氧化铜（$CuO$）、氧化亚铜（$Cu_2O$）、氧化钴（$CoO$）和氧化锰（$MnO$）等氧化物。

### （四）其他成分

釉中还含有三氧化二铝（$Al_2O_3$）和其他成分。$Al_2O_3$是一种重要成分，适量的$Al_2O_3$能与$SiO_2$形成网络，可增强釉层的强度及化学稳定性。

## 三、釉的分类

### （一）高温釉

#### 1. 高温釉概念

高温釉是1150°C以上烧成的釉。

#### 2. 高温釉出现

我国高温釉陶瓷的生产有着悠久的历史，最早的有实物考证的高温釉陶瓷是在商代遗址出土的陶器，上面施有一层薄薄的浅黄色釉或青灰色釉[60]。

#### 3. 高温釉化学成分

高温釉的化学成分主要有：$SiO_2$、$Al_2O_3$、$Fe_2O_3$、$TiO_2$、$CaO$、$MgO$、$K_2O$、$Na_2O$、$MnO$、$P_2O_5$等。其中$SiO_2$为网络形成剂，$CaO$、$MgO$、$K_2O$、$Na_2O$为助熔剂，$Fe_2O_3$、$MnO$为呈色剂。它们在釉中的含量参见表4-1-1[61]。

大量化学分析和有关统计表明，我国历代传统的高温釉都是在此基础上发展而来的[62]。

为了方便描述，将高温釉中影响其釉层结构的主要成分归纳成一个系统。根据它们的化学成分可将高温釉归于$Na_2O$-$K_2O$-$CaO$-$MgO$-$Al_2O_3$-$SiO_2$系统。

图4-1-2为中国历代青瓷釉的助熔剂组成，它反映了中国历代青瓷釉化学成分的演变规律，其他品种的高温釉大致也是按这样的规律演变而来的[63]。

### （二）低温釉

#### 1. 低温釉概念

低温釉是800°C－900°C烧成的釉。

#### 2. 低温釉出现

低温釉的发明比高温釉要晚得多，但在汉代已相当普遍。由于采用氧化铅（$PbO$）为主要助熔剂，所以也叫铅釉[64]。

# Section 1:
# Chemical Composition of Ceramic Glaze

## 1. The Definition of Glaze

A specifically mixed set of natural and mineral raw materials for application to the surface of a ceramic body that when fired, melts and then cools to become a glass phase layer over the body [56].

## 2. The Components of Glaze

Glaze primarily consists of: Networking Agents, Flux, and Coloring Agents.

### A. Networking Agents

Glass is the main phase of the glaze level; composed primarily of Silicone Dioxide ($SiO_2$) in tetrahedron shapes, mutually interlocking to create an irregular network; thus being slated as networking agents [57].

### B. Flux

During the melting of glaze, these solvents promote high temperature split reactions and accelerate the splitting of high-melting-point crystals (such as $SiO_2$) into eutectic solutions. Fluxes also bring rise to changes in the physical and chemical properties of the glaze (Such as mechanics, coefficient of expansion, viscosity, and chemical stability) [58].

Commonly used fluxes are: Sodium Oxide ($Na_2O$), Potassium Oxide ($K_2O$), Calcium Oxide ($CaO$), Magnesium Oxide ($MgO$), and Lead Oxide ($PbO$).

### C. Coloring Agents

Coloring Agents are substances between the clay body and glaze that act to add color to a ceramic piece [59]. They include: Iron Oxide ($FeO$), Ferric Oxide ($Fe_2O_3$), Cuprous Oxide ($CuO$), Cupric Oxide ($Cu_2O$), Cobaltous Oxide ($CoO$), Manganese Oxide ($MnO$), and other various metal oxides.

### D. Other Components

Glaze may contain trace amounts of Aluminum Oxide ($Al_2O_3$) and other trace elements. Aluminum Oxide is a more instrumental component in that, with the proper quantities, it can network with Silicone Dioxide to increase the strength and chemical stability of the glaze level.

## 3. Types of Glaze

### A. High-Temperature Glaze

#### I. High-Temperature Glaze Defined

Glazes that are formed in temperatures over 1150 °C (2100 °F) are considered High-Temperature.

#### II. The Induction of High-Temperature Glaze

The production of high-temperature glazes within China has a very long history. The earliest documented objects to bear high-temperature glazes are ceramics excavated from a Shang Dynasty ($17^{th}$-$11^{th}$ Century B.C.) site. They presented with a very thin layer of a pale yellow or graphite color glaze [60].

表 4-1-1 原始瓷釉的化学成分（%）

Table 4-1-1 Primitive Porcelain Glaze Chemical Compositions (%)

| 编号 # | 名称 Glaze Type | 出土地点 Excavation Location | 时代 Period | 化学组成 Chemical Composition | | | | | | | | | 总量 Total |
|---|---|---|---|---|---|---|---|---|---|---|---|---|---|
| | | | | SiO₂ | Al₂O₃ | Fe₂O₃ | TiO₂ | CaO | MgO | K₂O | Na₂O | MnO | P₂O₅ | |
| Sh20 | 原始瓷青灰色釉 Ashen Glaze | 河南郑州 Henan Zhengzhou | 商 Shang | 60.79 | 16.89 | 4.42 | 0.94 | 10.60 | 2.13 | 2.37 | 0.26 | 0.41 | | 98.81 |
| HZh1 | 原始瓷青灰色釉 Ashen Glaze | 河南郑州 Henan Zhengzhou | 商 Shang | 62.85 | 12.56 | 2.13 | 0.78 | 12.70 | 2.38 | 3.84 | 0.46 | 0.29 | 0.69 | |
| Sh16 | 原始瓷青灰色釉 Ashen Glaze | 河南郑州 Henan Zhengzhou | 商 Shang | 58.96 | 15.47 | 1.66 | 0.62 | 13.06 | 2.01 | 4.75 | 1.07 | 0.44 | | 98.04 |
| M216 | 原始瓷青釉 Celadon | 河南洛阳 Henan Luoyang | 西周 Western Zhou | 64.74 | 16.60 | 1.94 | | 9.51 | 1.84 | 3.82 | 0.28 | 0.38 | | 99.11 |
| HZH2 | 原始瓷青黄色釉 Celadon Yellow | 河南洛阳 Henan Luoyang | 西周 Western Zhou | 48.38 | 12.40 | 2.52 | 0.41 | 22.73 | 2.38 | 2.89 | 0.73 | 0.26 | 1.31 | |
| M-668 | 原始瓷青绿色釉 Celadon Green | 河南洛阳 Henan Luoyang | 西周 Western Zhou | | 13.49 | 3.06 | 1.41 | 18.42 | 3.86 | 2.98 | 0.81 | 0.52 | | |
| M-37 | 原始瓷青绿色釉 Celadon Green | 河南洛阳 Henan Luoyang | 西周 Western Zhou | | 14.53 | 3.80 | 1.41 | 17.94 | | 2.84 | 1.13 | 0.76 | | |
| M-250 | 原始瓷青绿色釉 Celadon Green | 河南洛阳 Henan Luoyang | 西周 Western Zhou | | 17.08 | 3.76 | 1.43 | 16.73 | 0.94 | 3.25 | 1.01 | 0.93 | | |
| M198 | 原始瓷青釉 Celadon | 河南洛阳 Henan Luoyang | 西周 Western Zhou | | 14.80 | 2.26 | 1.57 | 12.95 | 2.09 | 2.90 | 0.52 | 0.54 | | |
| M-325 | 原始瓷青釉 Celadon | 河南洛阳 Henan Luoyang | 西周 Western Zhou | | 15.68 | 2.03 | 1.37 | 11.75 | 2.28 | 3.80 | 1.07 | 0.58 | | |
| Zh10 | 原始瓷青灰色釉 Ashen Glaze | 浙江德清 Zhejiang Deqing | 西周 Western Zhou | | 11.71 | 3.35 | 0.73 | 9.93 | 2.47 | 5.11 | 1.34 | 0.24 | | |
| ZhJ4(1) | 原始瓷青釉 Celadon | 浙江江山 Zhejiang Jiangshan | 西周 Western Zhou | 67.57 | 15.61 | 2.07 | 0.57 | 7.23 | 1.36 | 3.17 | 0.30 | 0.29 | | |
| ZhJ4(2) | 原始瓷青釉 Celadon | 浙江江山 Zhejiang Jiangshan | 西周 Western Zhou | 66.26 | 15.05 | 1.84 | 0.97 | 10.07 | 1.62 | 3.74 | 0.42 | 0.52 | | 100.49 |
| ZhJ4(3) | 原始瓷青釉 Celadon | 浙江江山 Zhejiang Jiangshan | 西周 Western Zhou | 54.35 | 21.44 | 2.23 | 0.73 | 13.86 | 2.12 | 3.70 | 0.57 | 0.39 | | |
| ZhJ5 | 原始瓷青釉 Celadon | 浙江江山 Zhejiang Jiangshan | 西周 Western Zhou | 61.08 | 19.35 | 3.11 | 0.91 | 7.33 | 0.95 | 3.30 | 0.70 | 0.02 | | |

### III. The Chemical Composition of High-Temperature Glaze

The primary chemical components of high-temperature glazes are: $SiO_2$, $Al_2O_3$, $Fe_2O_3$, $TiO_2$, $CaO$, $MgO$, $K_2O$, $Na_2O$, $MnO$, and $P_2O_5$; with $SiO_2$, or Silica, acting as the networking agent; $CaO$, $MgO$, $K_2O$, $Na_2O$ as flux; and $Fe_2O_3$, $MnO$ as coloring agents (see table 4-1-1) [61]:

According to extensive chemical analysis, traditionally throughout Chinese history, high-temperature glazes have developed upon the foundation of those aforementioned [62].

For ease of use, we systemize high-temperature glazes based upon the primary elements that are acting on the composition of the glaze level. Therefore, the system for categorizing high-temperature glazes is based on a chemical composition of the $Na_2O$-$K_2O$-$CaO$-$MgO$-$Al_2O_3$-$SiO_2$ system.

Table 4-1-2 outlines the amount of flux composition among Chinese celadon glazes throughout history, and also shows the changes in chemical composition of celadon glazes over time [63].

## B. Low-Temperature Glaze

### I. Low-Temperature Glaze Defined

Glazes that are formed in temperatures between 800 and 900 °C (1470 °F-1650 °F) are considered Low-Temperature.

### II. The Induction of Low-Temperature Glaze

The invention of low-temperature glazes came much later than those of high-temperature glazes, but became very commonly used during the Han Dynasty (206 B.C.- 220 A.D.). Utilizing Lead Oxide (PbO) as the main flux, it earned the name Lead Glaze [64].

### III. The Chemical Composition of Low-Temperature Glaze

Table 4-1-3 shows a portion of low-temperature glaze chemical compositions [65]. According to their main components we can categorize them into $PbO$-$Al_2O_3$-$SiO_2$ and $PbO$-$SiO_2$ systems.

## C. Medium-Temperature Glaze

### I. Medium-Temperature Glaze Defined

Glazes that are formed in the temperature range around 1000 °C (1830 °F) are considered Medium-Temperature.

### II. The Induction of Medium-Temperature Glaze

Before the Song Dynasty (960-1279 A.D.), there were only two types of glaze: High-temperature glaze and low-temperature glaze.

However, during the Song Dynasty, the Cizhou Kiln invented a new medium-temperature glaze called Peacock-Green Glaze. With the flux consisting primarily of the Alkaline Metal Oxide ($K_2O$), it is also called Alkaline Glaze. Cizhou's peacock-green glaze was the first alkaline glaze in China [66].

### III. The Chemical Composition of Medium-Temperature Glaze

Table 4-1-4 shows a portion of medium-temperature glaze chemical compositions from a selection of Ming Dynasty Xuande Period (1426-1435 A.D.) peacock-green glaze vessels [67]. According to their main components we can categorize the majority into $K_2O$-$SiO_2$ and $K(Na)_2O$-$SiO_2$ systems, and only a small minority into the $K_2O$-$PbO$-$SiO_2$ or $K_2O$-$CaO$-$SiO_2$ systems.

## D. Summary

Traditional Chinese Ceramic Glazes can be broken into three Primary Categories:

I. High-Temperature, fired over 1150 °C, $Na_2O$-$K_2O$-$CaO$-$MgO$-$Al_2O_3$-$SiO_2$ system.

II. Low-Temperature, fired at 800 °C-900 °C, $PbO$-$Al_2O_3$-$SiO_2$ or $PbO$-$SiO_2$ system.

III. Medium-Temperature, fired at approximately 1000 °C, $K(Na)_2O$-$SiO_2$ or $K_2O$-$PbO$-$SiO_2$ and $K_2O$-$CaO$-$SiO_2$ system.

During different periods, these three types of glaze

## 3. 低温釉化学成分

表4-1-3是部分低温釉的化学成分[65]。根据它们的化学成分，可将低温釉分成PbO-$Al_2O_3$-$SiO_2$和PbO-$SiO_2$系统。

## （三）中温釉

### 1. 中温釉概念

中温釉是1000°C左右烧成的

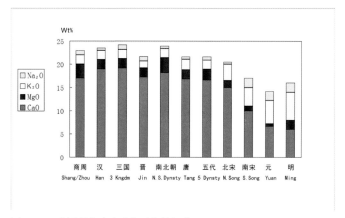

图4-1-2 中国历代青瓷釉的助熔剂组成
Table 4-1-2 Amount of Flux Composition Amongst Chinese Celadon Glazes Throughout History

### 表 4-1-3 唐三彩、辽三彩、红绿彩、琉璃釉、娇黄釉的化学组成
Table 4-1-3
Tang Three-Color, Liao Three-Color, Red and Green Color, Colored Glaze, and Yellow Glaze Chemical Compositions

| | $SiO_2$ | $Al_2O_3$ | $Fe_2O_3$ | CaO | MgO | $K_2O$ | $Na_2O$ | PbO | CuO |
|---|---|---|---|---|---|---|---|---|---|
| 巩县唐三彩绿釉 Tang 3 Color Green | 30.66 | 6.56 | 0.56 | 0.88 | 0.25 | 0.79 | 0.36 | 49.77 | 3.81 |
| 巩县唐三彩黄釉 Tang 3 Color Yellow | 28.65 | 8.05 | 4.09 | 1.65 | 0.42 | 0.72 | 0.45 | 54.59 | |
| 巩县唐三彩蓝釉 Tang 3 Color Blue | | | 0.99 | 0.79 | 0.43 | 0.88 | 0.22 | 45.00 | |
| 宋绿釉 Song Green Glaze | 32.26 | 4.83 | 1.41 | 2.24 | 0.47 | 0.65 | 0.31 | 54.64 | 2.80 |
| 辽三彩黄釉 Liao 3 Color Yellow | 33.05 | 6.58 | 3.53 | 1.01 | 0.14 | 0.67 | 0.39 | 53.67 | |
| 辽三彩绿釉 Liao 3 Color Green | 33.28 | 3.45 | 1.48 | 0.93 | 0.12 | 0.49 | 0.24 | 57.22 | 2.09 |
| 磁州窑绿彩 Cizhou Green | 40.56 | 6.56 | 0.36 | 6.59 | 0.39 | 0.34 | | 43.36 | 1.84 |
| 磁州窑红彩 Cizhou Red | 39.31 | 7.10 | >40 | 5.42 | 0.00 | 2.32 | 0.23 | 2.88 | 0.36 |
| 元代琉璃瓦绿釉 Yuan Green | 34.22 | 4.25 | 0.32 | 0.51 | 0.08 | 1.90 | 0.17 | 56.88 | 2.28 |
| 明代琉璃瓦绿釉1 Ming Green 1 | 30.35 | 5.23 | 0.17 | 0.40 | 0.13 | 0.18 | 0.74 | 60.04 | 3.46 |
| 明代琉璃瓦绿釉2 Ming Green 2 | 45.51 | 3.52 | 0.37 | 0.66 | 0.14 | 7.38 | 4.27 | 34.73 | 3.35 |
| 清代琉璃蓝釉 Qing Blue | 69.63 | 4.51 | 0.73 | 11.35 | 1.42 | 0.72 | 9.90 | 1.35 | |
| 清代琉璃瓦黄釉 Qing Yellow | 35.74 | 5.48 | 2.98 | 0.46 | 0.23 | 0.67 | 0.39 | 53.57 | |
| 弘治娇黄釉 Hongzhi Yellow | 42.93 | 4.52 | 3.66 | 1.16 | 0.10 | 1.30 | 0.73 | 45.00 | 0.05 |

唐三彩蓝釉含有1.03% CoO，清代琉璃蓝釉的CoO未测。

compositions underwent different changes.

# 4. Formulation of Glaze

## A. High-Temperature Glaze Formulation

Traditional Chinese high-temperature glazes were primarily formulated using: Porcelain clay and added plant ash or glaze ash [68].

The CaO found in high-temperature glaze comes primarily from plant ash or glaze ash. The high-temperature glazes fired before the Southern Song Dynasty contained a relatively high proportion of CaO, with $Na_2O$ and $K_2O$ being relatively low. While the glazes after Southern Song Dynasty used a higher amount of porcelain clay and less ash, creating a mixture with relatively low CaO and high $Na_2O$ and $K_2O$ [69].

Luo et al put forth that glazes utilizing plant ash as the sample selection showed that the primary flux was CaO (thus named Calcium Glazes). These Calcium Glazes can be further categorized in Calcium Glaze, Calcium Alkaline Glaze, and Alkaline Calcium Glaze [70].

## B. Low-Temperature Glaze Formulation

Traditional Chinese low-temperature glazes were primarily formulated using: Quartz powder, lead powder, and coloring agents [71].

## C. Medium-Temperature Glaze Formulation

Traditional Chinese medium-temperature glazes were primarily formulated using: Quartz powder, Potassium Nitrate, and coloring agents or quartz powder, Potassium Nitrate, coloring agents, and Lead powder [72].

表 4-1-4 景德镇御窑宣德孔雀绿釉瓷片（C22）化学组成（LA-ICP-AES测定）
Table 4-1-4 Jingdezhen Imperial Kiln Xuande Peacock Green Glaze (PGG) Shard (C22) Chemical Composition

宣德孔雀绿釉瓷片（C22）化学组成（重量%）
Xuande Peacock Green Glaze Shard (C22) Chemical Composition (%)

| 类别 Type | $SiO_2$ | $Al_2O_3$ | $K_2O$ | $Na_2O$ | CaO | MgO | CuO | $Fe_2O_3$ | PbO | $P_2O_5$ | $SnO_2$ | $TiO_2$ | ZnO |
|---|---|---|---|---|---|---|---|---|---|---|---|---|---|
| 孔雀绿釉-1 PGG-1 | 73.48 | 1.32 | 7.62 | 4.54 | 0.24 | 0.14 | 9.65 | 0.37 | 1.19 | 0.61 | 0.65 | 0.02 | 0.04 |
| 孔雀绿釉-2 PGG-2 | 71.20 | 2.10 | 10.04 | 6.25 | 0.18 | 0.13 | 7.30 | 0.31 | 1.19 | 0.46 | 0.66 | 0.02 | 0.03 |

宣德孔雀绿（C22）白釉、胎及化妆土化学组成（重量%）PGG C22 White Glaze, Body, and Slip Chemical Composition (%)

| 类别 Type | $SiO_2$ | $Al_2O_3$ | $K_2O$ | $Na_2O$ | CaO | MgO | CuO | $Fe_2O_3$ | PbO | $P_2O_5$ | MnO | $TiO_2$ | ZnO |
|---|---|---|---|---|---|---|---|---|---|---|---|---|---|
| 白釉 White | 72.56 | 14.85 | 5.03 | 1.65 | 4.29 | 0.19 | 0.01 | 0.80 | 0.08 | 0.28 | 0.06 | 0.03 | 0.01 |
| 化妆土 Slip | 57.88 | 34.49 | 3.94 | 1.78 | 0.05 | 0.32 | 0.07 | 0.93 | 0.04 | 0.17 | 0.04 | 0.10 | 0.03 |
| 胎 Body | 68.23 | 26.47 | 3.28 | 0.61 | 0.04 | 0.19 | 0.01 | 0.69 | 0.05 | 0.17 | 0.02 | 0.07 | 0.00 |

注：1）孔雀绿釉-1和孔雀绿釉-2为从釉表面到釉内部剥蚀过程中两次测定结果，分别反映釉表面和釉内部的组成情况。
2）对附在胎上的化妆土直接进行剥蚀，故测定的化妆土组成可能会受胎的影响，但可反映出其与胎组成的不同。

釉。

### 2. 中温釉出现

宋代以前的陶瓷釉只有高温釉和低温釉两种。

宋代磁州窑发现的孔雀绿釉，就是中温釉，它以碱金属氧化物$K_2O$为主要助熔剂，也称为高碱釉。磁州窑孔雀绿釉是我国最早的高碱釉[66]。

### 3. 中温釉化学成分

表4-1-4是明代宣德孔雀绿釉的化学成分[67]。根据它们的化学成分，可将中温釉分成$K_2O$-$SiO_2$系统和K(Na)$_2$O-$SiO_2$系统、还有少数属于$K_2O$-PbO-$SiO_2$系统和$K_2O$-CaO-$SiO_2$系统。

### （四）小结

我国传统的陶瓷釉按照烧制温度可分为三类：

1. 高温釉。烧成温度在1150°C以上，属于$Na_2O$-$K_2O$-CaO-MgO-$Al_2O_3$-$SiO_2$系统。

2. 低温釉。烧成温度在800°C-900°C，属于PbO-$Al_2O_3$-$SiO_2$或PbO-$SiO_2$系统。

3. 中温釉。烧成温度在1000°C左右，属于K(Na)$_2$O-$SiO_2$、$K_2O$-PbO-$SiO_2$或$K_2O$-CaO-$SiO_2$系统。

在不同的时期，这三种陶瓷釉的各组分含量有不同的变化。

## 四、釉的配方

### （一）高温釉配方

中国历代传统的高温釉主要是由瓷石加草木灰或釉灰配制而成[68]。

高温釉中的CaO主要由草木灰或釉灰提供，南宋以前的高温釉极大部分都是CaO含量较高，$Na_2O$、$K_2O$含量相对较低；南宋以后增加了瓷石的用量，减少了釉灰的用量，所以CaO含量大幅度降低，而$Na_2O$、$K_2O$含量大大提高[69]。

罗宏杰等人提出，可用草木灰的釉式统计值作为参照标准来划分以CaO为主要助熔剂的瓷釉（钙系釉）。所以钙系釉可分为钙釉、钙碱釉和碱钙釉[70]。

### （二）低温釉配方

中国历代传统的低温釉主要是由石末、铅粉以及着色剂配制而成[71]。

### （三）中温釉配方

中国历代传统的中温釉主要是由石末、硝、着色剂或石末、硝、着色剂及铅粉配制而成[72]。

# 第二节
# 陶瓷釉化学成分的作用

从第一节我们了解到釉层是各类氧化物在高温作用下被熔化，然后冷却形成的玻璃体。但并不是所有的氧化物都能形成玻璃体，只有在氧化物的分子结构能形成特定的网络结构时，才能形成玻璃体。

## 一、网络形成剂作用

网络形成剂主要由$SiO_2$构成。$SiO_2$的基本结构是硅氧四面体（简写为$[SiO_4]$四面体），Si与O通过硅氧键（Si—O）相连，一个硅原子与四个氧原子相连接形成四面体（见图4-2-1）。为了便于书写，将其简化成≡Si—。

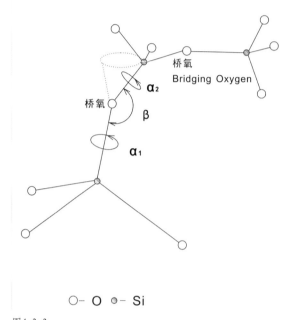

图4-2-2

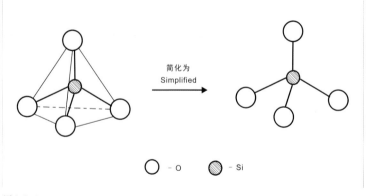

图4-2-1

氧原子通过与两个硅原子相连将四面体连接起来，连接这四面体的氧原子称为桥氧。它们以顶角相连而构成连续的架状三维硅氧网络（见图4-2-2[73]）。为了便于书写，将其简化成≡Si—O—Si≡（中间的O为桥氧）。

≡Si—O—Si≡的连接角度是不定的，也就是说，图4-2-2中的$α_1$、$α_2$和$β$角的度数是可变化的，故就整体而言，形成的是不规则的硅氧网络。通过这样的连接形成的$SiO_2$三维网络（也就是玻璃体），是没有对称性和周期性的，它是最简单的单组份硅酸盐玻璃。

由于$SiO_2$是一种能相互连接形成三维空间网络的氧化物，我们称之为网络形成体氧化物。$SiO_2$是釉层的主要成分，机械强度及化学稳定性都比较好，所以釉层有保护陶瓷的作用。

## 二、助熔剂作用

### （一）高温釉、中温釉助熔剂

釉料配方中的石英熔点较高，达到1750°C，而当代瓷窑最高的烧成温度也不过1400°C。由第一节可

# Section 2:
# The Functions of Ceramic Glaze Chemical Components

According to Section 1, glaze is a fusion of certain oxidized components melded under high temperatures, and cooled to form a glass-like coating. However, not all oxidized compounds have the ability to form glass; only those that have set networking construction are able to arrange as such.

## 1. The Function of Networking Agents

Networking agents are comprised primarily of Silica $SiO_2$. The foundational construction of $SiO_2$ is a Silicone and Oxygen tetrahedron (Simplified as $[SiO_4]$ tetrahedron) in which one Si atom is combined with four O atoms to form the shape illustrated in Picture 4-2-1. Hereafter to be simplified as $\equiv Si-$.

An Oxygen atom that links two Silicone atoms is called a 'bridging Oxygen'. They affix in static angles to form a three-dimensional silica network structure illustrated in 4-2-2 [73]. Hereafter to be simplified as $\equiv Si-O-Si\equiv$ (the central O as the bridging Oxygen).

Because of the unfixed angle of attachment of the $\equiv Si-O-Si\equiv$ we can see in Picture 4-2-2 that angles $a_1$、$a_2$ and $\beta$ can vary. It is due to this that $SiO_2$ molecules fuse to form an irregularly shaped network structure that lacks a symmetry and periodicity; this structure is generically called silicate glass. This irregular construction of $SiO_2$ is the main building block of ceramic glaze.

Because $SiO_2$ is able to interconnect and form a three-dimensional network structure, we call it a networking Oxide. This glaze level geometric structure is relatively strong with a good chemical stability, creating an inherent functionality of protecting the ceramic body.

## 2. The Function of Flux

### A. High and Medium-Temperature Flux

Quartz, used in glaze, has a very high melting point, as high as 1750°C; but currently the highest temperature used in firing ceramics is around 1400°C. Because of this, alkaline metal oxides $Na_2O$ and $K_2O$ as well as alkaline earth metal oxides $CaO$ and $MgO$ are added to high and medium-temperature glazes to act as a flux, effectively lowering the melting temperature of quartz and allowing for a lower firing temperature.

During the fusion of $SiO_2$, if alkaline metal oxides ($R_2O$, where R represents Na, K) or alkaline earth metal oxides (RO where R represents Ca, Mg) are added, the bond of the R—O is primary and much weaker than that of Si—O. This is to say that the Si—O $Si^{4+}$ ions can attract the $O^{2-}$ ions of the R—O creating a bond of $\equiv Si-O-R$ that has a $\equiv Si-O$ bond that is stronger than that of the $\equiv Si-O$ bond found in $\equiv Si-O-Si\equiv$. The result is the breaking away of the bridging O ($\equiv Si\cdots O-Si\equiv$). Thus, the bridging Oxygen becomes a non-bridging Oxygen and a new $\equiv Si-O-R$ bond is formed. After the breakdown of $\equiv Si-O-Si\equiv$, the $\equiv Si-O$ bond's strength, length, and angle all undergo change [74] (see picture 4-2-3).

Therefore, the network experiences several locations of breakdown where Si and O are no longer conjoined on both sides, but rather an O and R bond takes its place.

$SiO_2$ silicone dioxide has a Si:O ratio of 1:2. With the addition of $R_2O$ or RO, this Si:O ratio can increase to 1:3 or 1:4, from which the three-dimensional $[SiO_4]$ tetrahedron network is split into stratified, chain, and toroidal structures until the entire network is spilt into island structured $[SiO_4]$ tetrahedrons. During the breaking down process, chains of

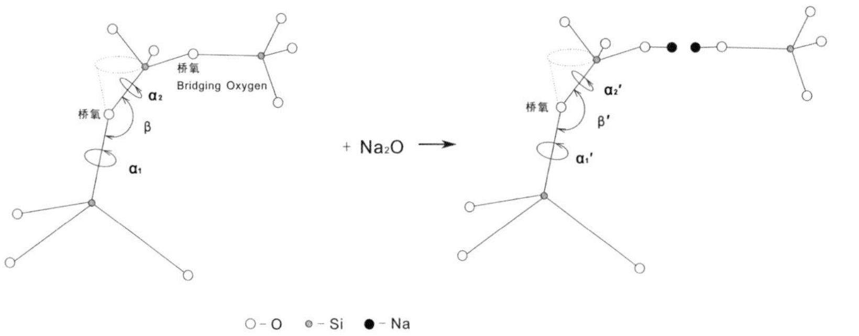

图4-2-3

知，碱金属氧化物$Na_2O$和$K_2O$以及碱土金属氧化物CaO、MgO在陶瓷高温釉和中温釉中都可充当助熔剂的角色，它们可以降低石英的熔点，从而降低陶瓷的烧结温度。

当在$SiO_2$熔体中加入碱金属（$R_2O$，R在此指Na、K）或碱土金属氧化物（RO，R在此指Ca、Mg）时，由于R—O是离子键为主，其键强比Si—O键弱得多，因此，Si—O键中的$Si^{4+}$能把R—O键上的$O^{2-}$吸引到自己的周围，使≡Si—O—R中的≡Si—O键比≡Si—O—Si≡中的≡Si—O键还要强，结果是使原来的桥氧键（≡Si⋯O—Si≡）从虚线处断开，桥氧成为非桥氧，同时形成新的≡Si—O—R键。≡Si—O—Si≡断裂后，Si—O键的键强、键长和键角都会发生变化[74]（见图4-2-3）。

这样，硅氧网络中许多地方会出现断点，即O不是与两个Si相连，而是其中一端与R相连。

$SiO_2$的硅氧比（Si/O）为1∶2。随着$Na_2O$、$K_2O$、CaO的加入，Si/O可由原来的1∶2上升到1∶3或1∶4，从而使[$SiO_4$]四面体的架状三维网络发生断裂，连接方式随着硅氧比的不同会由架状变为层状、链状、环状，直至最后桥氧全部断裂形成岛状的[$SiO_4$]四面体。在这个过程中，许多断裂的[$SiO_4$]四面体短链会聚集，形成低聚物。这种低聚物不断增多，在一定条件下会产生相互作用，形成高聚物，直至平衡。所以，釉层的结构会因$Na_2O$、$K_2O$、CaO、MgO的加入而发生变化。

这些碱金属氧化物（$Na_2O$、$K_2O$）和碱土金属氧化物（CaO、MgO）的加入，并没有参与到硅氧网络中去，而是使网络断裂，从而使其熔点降低。它们改变了硅氧网络的性质，我们称它们为网络外体氧化物。〔注：MgO只有在碱金属氧化物较多而且不存在$Al_2O_3$及$B_2O_3$等氧化物时，才可进入硅氧网络。〕

## （二）低温釉助熔剂

由第一节可知，PbO在陶瓷釉中可充当助熔剂的角色。

根据PbO的X射线结构分析，晶态PbO的结构如图所示（见4-2-4）。由图可以看出，$Pb^{2+}$被8个$O^{2-}$包围，其中4个$O^{2-}$距离$Pb^{2+}$较远（0.429nm），另外4个$O^{2-}$距离$Pb^{2+}$较近（0.230nm），形成不对称配位。$Pb^{2+}$外层的惰性电子受较近的4个$O^{2-}$的排斥，被推向另外4个$O^{2-}$的一侧，因此在晶态PbO中组成一种四方锥体[$PbO_4$]的结构单元[75]。

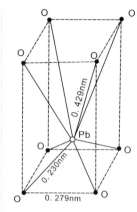

正方形PbO原子间距示意图
PbO Interatomic Distance Schematic

图4-2-4

由图4-2-5可看出，$Pb^{2+}$处于四方锥体的顶端，$Pb^{2+}$惰性电子处于远离4个$O^{2-}$的一面。一般来说，在铅含量较高的玻璃中均存在这种四方锥体，它形成一种螺旋形的链状结构，在玻璃中与[$SiO_4$]四面体通过顶角或共边相连接，形成一种特殊的网络（见图4-2-6）。但当玻璃中的PbO含量较高时，[$SiO_4$]四面体不能提供足够的桥氧与$Pb^{2+}$形成网络，所以有一部分$Pb^{2+}$只能作为网络外体存在于玻璃中。正是这种特殊的网络，使$PbO-SiO_2$系统具有很宽的玻璃形成区，决定了PbO在硅酸盐熔体中的高度助熔性[76]，能使$SiO_2$熔点降低，降低烧结温度。

PbO既可以与$SiO_2$形成网络，也可处于网络外，所以我们把它们称为中间体氧化物。

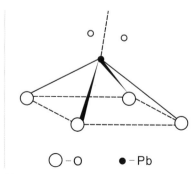

图4-2-5

[SiO_4] tetrahedrons come together to form low polymers, under the right circumstances these low polymers begin to interact and form high polymers until equilibrium is reached. Therefore, because of the addition of $Na_2O$, $K_2O$, CaO, MgO, the structure of the glaze undergoes change.

These alkaline metal oxides and alkaline earth metal oxides do not enter into the silicate networking, but only assist in the network breakdown and bringing down the melting point. They only work to change the structural characteristics of the silicate network, so we call them external oxides (note: Without the presence of $Al_2O_3$, $B_2O_3$ or other Oxides, and with large amounts of alkaline metal oxides present, MgO can enter into the silica network).

## B. Low-Temperature Flux

According to Section 1, we know that PbO can act as a glaze flux.

Looking at the microstructure of the PbO crystal in Picture 4-2-4 we can see that the $Pb^{2+}$ ion becomes lopsidedly encompassed by 8 $O^{2-}$ ions, 4 located at a further distance and 4 located at a closer distance (.429 and .230nm respectively). The force of the $Pb^{2+}$ ion's outer electrons receive repelling force from four $O^{2-}$ ions that are closer to them, pushing towards the other four $O^{2-}$ ions and creating the pyramid shaped unit [PbO_4] compound illustrated [75].

As Picture 4-2-5 shows, the $Pb^{2+}$ ion is located at the pinnacle of the pyramid shape while the negatively charged $O^{2-}$ ions make up the base of the structure. Generally, within a glass that has a considerably high concentration of lead, a helix chain is formed by the convergence of the pyramid in conjunction with the vertex angle or side of the [SiO_4] tetrahedron (see picture 4-2-6). When the amount of PbO is too high, there is not enough [SiO_4] tetrahedron bridging Oxygen to bond with the $Pb^{2+}$ ions, so some of the remainder $Pb^{2+}$ ions remain outside the network within the glass. It is precisely this specialized networking capability that allows the $PbO-SiO_2$ system of glazes to occur at lower temperatures [76]. The PbO within the Silicate acts as the flux which brings down the melting point of $SiO_2$ to a lower level.

PbO can network together with $SiO_2$ as well as remain outside the glass network; because of this we call them intermediate oxides.

## 3. The Function of Other Components

Apart from networking agents, flux, and coloring agents; high-temperature glaze contains a few other inclusions, such as Aluminum Oxide ($Al_2O_3$).

Within silicate glass, positive $Al^{3+}$ ions of $Al_2O_3$ have two types of pairing; 4 and 6 arranged into tetrahedron and hexahedron shapes respectively. When the Moore Ratio $R_2O/Al_2O_3>1$, $Al^{3+}$ ions match to the [SiO_4] tetrahedron as a [AlO_4] tetrahedron network. This tightly woven network improves the stability of the silica glass. However, when the Moore Ratio $R_2O/Al_2O_3<1$, the amount of alkaline metal ions are less than the $Al^{3+}$ ions, the $Al^{3+}$ ions exceed chemical equilibrium and are pushed into hexahedron pairing with the silicate glass network rather than the tetrahedron shape [77]. This creates a more loosely woven network with lower density of silicate glass and thus a lower physical chemical stability [78].

$Al_2O_3$ can network together with $SiO_2$ as well as remain outside the glass network; because of this we call them intermediate oxides as well.

## 4. The Relationship Between Glaze Composition and Glaze Stability

### A. High and Mid-Temperature Glazes

In order to maintain ionic neutrality, alkaline metal ions and alkaline earth metal ions will disperse evenly but without order throughout the naturally occurring spaces within the silica tetrahedron network of high and medium-temperature glazes. These ions and atoms that have weaker bonds with the Si—O can collect and move throughout the glaze level. For example: $H^+$, $Li^+$, $Na^+$, $K^+$, $Rb^+$, $Cs^+$, $Ag^+$, $Cu^+$, and other monovalent ions can network and enter into, without breaking down the glaze level. Even under normal temperatures, they can continue to move within the silicate glass. Additionally, small molecules such as Hydrogen and water can also move throughout the glaze level [79]. According to the above stated,

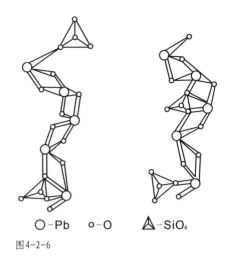

图 4-2-6

## 三、其他成分作用

高温釉中除了网络形成剂、助熔剂、着色剂以外，还有其他成分，如三氧化二铝（$Al_2O_3$）。

在硅酸盐玻璃中，$Al_2O_3$ 中的 $Al^{3+}$ 有 4、6 两种配位状态，即位于四面体或八面体中。当摩尔比 $R_2O/Al_2O_3>1$ 时，$Al^{3+}$ 位于四面体中，以 $[AlO_4]$ 四面体形式将一些 $[SiO_4]$ 四面体的断点连接起来，与 $[SiO_4]$ 组成统一的网络，使玻璃结构趋向紧密，对硅氧网络起补网作用，提高釉层结构的稳定性；当摩尔比 $R_2O/Al_2O_3<1$ 时，碱金属离子不足以使 $Al^{3+}$ 离子形成 $[AlO_4]$ 四面体时的化合价平衡，超出的 $Al^{3+}$ 离子只能在玻璃中以六配位的形式[77]，作为网络外体存在于八面体之中，从而使玻璃结构趋向于松散，使硅氧网络的紧密程度下降，化学稳定性也随之下降[78]。

$Al_2O_3$ 既可以与 $SiO_2$ 形成网络，也可处于网络外，所以它也是一种中间体氧化物。

## 四、釉的化学成分与釉层结构稳定性的关系

### （一）高温釉及中温釉

在高温釉及中温釉中，由于 $[SiO_4]$ 四面体间隙的存在，碱金属或碱土金属离子会均匀而无序地分布在某些 $[SiO_4]$ 四面体之间的空隙中，以维持网络中局部的电中性。与 Si—O 连接较弱的原子、离子或原子团在釉层中可以移动，如 $H^+$、$Li^+$、$Na^+$、$K^+$、$Rb^+$、$Cs^+$、$Ag^+$、$Cu^+$ 等一价离子可以不破坏釉层网络而进入釉层之中，即使在常温下，也较易在玻璃中移动；此外，$H_2$、$H_2O$ 等小分子也可以在釉层中通过[79]。由前述得知，碱金属氧化物的介入，打断了原有 $[SiO_4]$ 四面体的网络链接；同时，碱金属离子的移动，使釉层易于与空气、水、酸、碱发生反应，从而釉层遭到腐蚀。所以碱金属氧化物含量越高，釉层抗腐蚀能力越差，机械强度也会减弱。

此外，中间体氧化物 $Al_2O_3$ 的加入可使 $[SiO_4]$ 网络加强，釉层化学稳定性提高；但如果 $Al_2O_3$ 含量过高，反而会使网络紧密程度下降，化学稳定性也会下降[80]。

### （二）低温釉

与高温釉、中温釉中的碱金属离子类似，低温釉中的 $Pb^{2+}$ 离子不断的移动、迁出，使釉层易于与空气、水、酸、碱发生反应，从而釉层遭到腐蚀。所以 PbO 含量越高，釉层抗腐蚀能力越差，机械强度也会减弱。

低温釉的化学稳定性及机械强度都不如高温釉。

### （三）小结

不同种类釉层的化学成分可归纳成：

1. 高温釉的釉层是玻璃体（$Na_2O$-$K_2O$-$CaO$-$MgO$-$Al_2O_3$-$SiO_2$ 系统），它是由以 $[SiO_4]$ 四面体为基础结构的三维空间网络所组成的非晶体固体。

2. 中温釉的釉层是玻璃体（$K(Na)_2O$-$SiO_2$、$K_2O$-$PbO$-$SiO_2$ 系统或 $K_2O$-$CaO$-$SiO_2$ 系统），它是由以 $[SiO_4]$ 四面体为基础结构的三维空间网络所组成，或由 $[SiO_4]$ 四面体与 $[PbO_4]$ 四方锥体形成的螺旋形链状结构网络所组成的非晶体固体。

3. 低温釉的釉层是玻璃体（$PbO$-$SiO_2$ 或 $PbO$-$Al_2O_3$-$SiO_2$ 系统），它是由 $[SiO_4]$ 四面体与 $[PbO_4]$ 四方锥体形成的螺旋形链状结构网络所组成的非晶体固体。

高温釉釉层的抗风化能力与碱金属、碱土金属氧化物的含量有关，这些氧化物含量越高，釉层抗风化能力越差。高温釉中 $Na_2O$、$K_2O$ 含量较低，所以釉层结构较为稳定，抗风化能力较好。

中温釉釉层的抗风化能力与碱金属、碱土金属氧化物及 PbO 的含量有关，这些氧化物含量越高，釉层抗风化能力越差。中温釉以及低温釉的碱金属、碱土金属氧化物及 PbO 的含量都比较高，与高温釉相比，抗风化能力较差。

the introduction of alkaline metal oxides breaks down the tetrahedron [$SiO_4$] network. At the same time, the presence of alkaline metal ions increases the likelihood of the glaze level receiving corrosion caused by reaction with air, water, acids, and bases. Therefore, the higher the concentration of alkaline metal oxides within the glaze, the lower the resistance the glaze has to weathering and decay, as well as having a weakened mechanical strength.

Moreover, the addition of $Al_2O_3$ into the [$SiO_4$] network increases the strength and chemical stability of the glaze level. However, if the amount of $Al_2O_3$ becomes too high, the network density decreases and with it the strength and chemical stability [80].

## B. Low-Temperature Glazes

Similar to that of the alkaline metal ions of high and medium-temperature glaze, $Pb^{2+}$ ions within low-temperature glaze continually move and transfer out, which increases the likelihood of the glaze level receiving corrosion caused by reaction with air, water, acids, and bases. Therefore, the higher the concentration of lead within the glaze, the lower the resistance the glaze has to weathering and decay, as well as having a weakened mechanical strength.

Low-temperature glaze chemical stability and mechanical strength are not as high as that of high-temperature glazes.

## C. Summary

The Different types of glazes can be categorized according to composition as such:

I. High-temperature glaze level is in glass phase ($Na_2O$-$K_2O$-$CaO$-$MgO$-$SiO_2$-$Al_2O_3$), it is a non-crystalline solid formed on the foundation of a three-dimensional [$SiO_4$] tetrahedron network.

II. Medium-temperature glaze level is in glass phase ($K(Na)_2O$-$SiO_2$ or $K_2O$-$PbO$-$SiO_2$ and $K_2O$-$CaO$-$SiO_2$), it is a non-crystalline solid formed on the foundation of a three-dimensional [$SiO_4$] tetrahedrons and $PbO_4$ into helix chain pyramids.

III. Low-temperature glaze level is in glass phase ($PbO$-$SiO_2$ or $PbO$-$Al_2O_3$-$SiO_2$), it is a non-crystalline solid formed on the foundation of a three-dimensional network of [$SiO_4$] tetrahedrons and $PbO_4$ into helix chain pyramids.

The resistance to weathering of high-temperature glazes is related to the amount of alkaline metal and alkaline earth metal oxides within the glaze level, with the higher the amount of these oxides the lower the resistance. High-temperature glazes with relatively lower $Na_2O$ and $K_2O$ have a higher chemical stability and resistance to weathering.

The resistance to weathering of medium-temperature glazes is related to the amount of alkaline metal oxides, alkaline earth metal oxides, and PbO within the glaze level, with the higher the amount of these oxides the lower the resistance. Medium-temperature glaze as well as low-temperature glaze with relatively higher alkaline metal oxides, alkaline earth metal oxides, and PbO, compared to that of high-temperature glazes, have a lower resistance to weathering.

# 第三节
# 陶瓷釉表面活性

## 一、釉表面活性概念

釉表面活性是陶瓷釉表面与外部环境中的物质发生反应的能力。

## 二、釉表面活性产生原因

釉表面具有活性的产生原因主要有：

1. 釉料在高温熔融后形成玻璃体，由于高温与助熔剂的作用，玻璃熔体中会有很多断键，不完全配位。在釉料冷却时断键应该合拢，但由于釉表面冷却速度比内部快，断键来不及合拢，键角也不容易恢复，从而在釉表面存在着更多断键。因为釉表面的断键，釉表面会产生极性，从而会在大气中吸收羟基，这些羟基可以吸附水或与其他化合物反应。

2. 有研究指出：在玻璃熔体中，$Al_2O_3$挥发量为0.5%－5%，$B_2O_3$为1%－5%，$Na_2O$和$K_2O$的挥发量可达5%。釉表面成分的挥发会造成空位，这些空位迁移聚集，会在釉表面形成多个孔隙，使其具有多孔性[81]。

3. 釉裂纹的存在。

## 三、小结

陶瓷釉表面具有活性，易于吸附和积聚水份，从而便于釉表面的离子交换。所以陶瓷釉表面容易与酸、碱、盐溶液发生反应，造成釉表面的风化。

# Section 3:
# Active Nature of the Ceramic Glaze Surface

## 1. The Active Nature of the Glaze Surface Defined

This is the ability of the glaze surface and materials within the external environment to undergo physical and chemical changes.

## 2. Causes of the Glaze Surface Activity

Causes of activity on the Glaze surface are roughly split into three types:

A. After the high temperature firing of glaze, it forms a glass. Due to high temperature and the presence of flux, the glass body has many broken bonds and is not completely coordinated evenly. During the cooling of the glaze, these broken bonds should join together; but because the surface of the glaze cools faster than the center of the glaze, not all broken bonds are able to join together and the bonding angles are difficult to repair. Because of this, the glaze surface presents with numerous bond separations. Glaze surface bond separations gives rise to chemical polarity, which absorbs hydroxyls from the surrounding atmosphere. These hydroxyls absorb water and react with other chemical compounds, leaving the face of the glaze chemically activated.

B. Research shows: Within molten glass, with a volatized $Al_2O_3$ quantity of 0.5%-5% and $B_2O_3$ 1%-5%; volatized $Na_2O$ and $K_2O$ quantity can reach 5%. This volatized composition can create empty space, these empty spaces, in turn, migrate and assemble to create small holes which effectively bring about a porous glaze surface [81].

C. The presence of glaze cracks.

## 3. Summary

The glaze surface is active; prone to attracting and absorbing surrounding water that partakes in ion swapping across the surface of the glaze. These characteristics make the surface of the glaze susceptible to weathering due to the ease of reactions with acids, bases, and salt solutions.

# 第四节 陶瓷釉层形貌

## 一、釉层概念

釉层是由玻璃相、气相、晶相及少量原料残留颗粒构成的覆盖在陶瓷坯体上的玻璃体[82]。

## 二、釉层形貌

### （一）釉层形貌概念

釉层形貌是釉层在一定观察角度下的三维可视状况。

### （二）釉层形貌特征

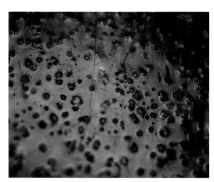

图4-4-1（T16）

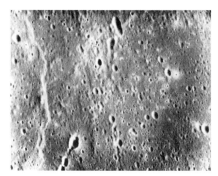

图4-4-2
美国国家航空航天局公布的水星表面照片
Image taken by NASA of the surface of Mercury

图4-4-3
阿波罗11号飞船在1969年绕月飞行时拍摄的月球背面环形山
Image taken in 1969 of the Moon by Apollo 11

用肉眼观察釉表面，非常平滑，但它在显微成像下却是凹凸不平（见图4-4-1），类似星球表面（见图4-4-2[83]、图4-4-3[84]）。

### （三）釉层形貌分类

#### 1. 玻璃相

（1）玻璃相概念

玻璃相是陶瓷高温烧制时各组成物质产生一系列物理变化、化学反应后冷却凝固形成的一种非晶态物质[85]。

（2）玻璃相先天形貌形成

陶瓷在烧制过程中产生气体（参见气相形成），当釉料熔化时，一部分气体从胎面（胎釉结合处）或釉层中逸出，一部分则留在釉层中形成气泡。由于气泡在熔融釉层中的运动而产生的扰动效应及胎面不平，烧制好的陶瓷釉表面会有凹凸不平的现象（见图4-4-4，图中箭头所指处为釉表面同一处在烧制过程中的变化）。

有些气孔由于降温较快或釉层粘度较大而没有闭合，从而使釉表面出现微小的孔洞（见图4-4-5）。不过这些孔洞过于微小，用500倍显微镜一般是观察不到的。

当气泡接近釉表面，温度下降时，气体体积缩小，待釉表面冷却凝固后，气泡上方的釉表面就会形成凹坑（见图4-4-6）。

如果气泡高出釉表面，温度下降时，气体体积缩小，高出釉表面的气泡上方会有凹坑出现，如同月球上的环形山（见图4-4-7），气泡周围较低的地方则形成沟壑。

#### 2. 气相

（1）气相概念

气相是存在于烧制好的陶瓷釉

# Section 4:
# Ceramic Glaze Level Appearance

## 1. Glaze Level Defined

Composed of glass, gas, crystal and trace amounts of other raw materials left over after the forming of glaze into a vitreous body. Essentially it is a layer of glass of varying thickness that covers all or a certain portion of the ceramic body [82].

## 2. Glaze Level Appearance

### A. Glaze Level Appearance Defined

Describes the three-dimensional state of the glaze surface; including the topography of the glaze.

### B. Glaze Level Appearance Characteristics

To the naked eye, glaze appears smooth and level; but under microscopic magnification the surface appears uneven with humps and pits (see picture 4-4-1), similar to the surface of certain celestial bodies (see pictures 4-4-2 [83] and 4-4-3 [84]).

### C. Glaze Level Appearance Types

I. Glass Phase

a. Glass Phase Defined

Refers to all the non-crystalline composite materials that emerged from a series of physical changes and chemical reactions during firing that solidify when cooled [85].

b. Glass Phase Formation

During the firing of ceramics, gaseous bodies are formed (see Gas Phase Formation). When the glaze is in a molten state, a portion of gasses from within the body or the glaze level escape, while the remainder remain within the glaze level. Due to the turbulent motion of the formation and movement of these air bubbles within the molten glaze level as well as the unevenness of the ceramic body, the post-firing glaze surface is topographically uneven. From illustration 4-4-4 we can see the progression of one point on the glaze surface as indicated by the red arrow.

There are some air holes, due to the rapid cooling or viscous nature of the glaze, that were unable to close and leave behind microscopic pits across the glaze surface (see picture 4-4-5). However, there are some pits that are microscopically invisible; while using a 500X magnification microscope, they are usually unobservable.

When air bubbles approach the surface while the glaze is cooling, the gas within the bubble contracts; this creates a small depression on the surface of the glaze over the top of the air pocket once the glaze hardens (see picture 4-4-6).

If an air bubble extrudes above the surface level while the glaze is cooling, the gas within the bubble still contracts; these bubbles create an extruding crater on the surface of the glaze (similar to that of moon craters, see picture 4-4-7). The area surrounding the air bubble is slightly lower, creating a trough.

II. Gas Phase

a. Gas Phase Defined

Gas phase is the gaseous bodies that are present within

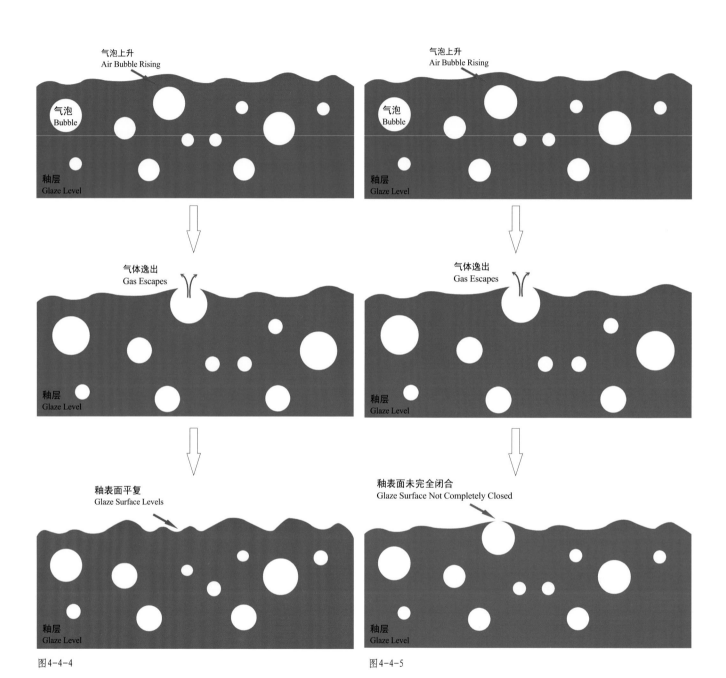

图 4-4-4　　　　　　　　　　　　　图 4-4-5

the glaze level after the firing of ceramics.

b. Gas Phase Formation

Gas phases are formed under the following circumstances [86]:

i. Ceramic Body-Glaze Reaction Formed Air Bubbles

The ceramic body presents with open and closed air pockets, when temperatures increase, these gaseous bodies expand. Open air pockets eject air directly into the glaze level. Molten glaze also moistens the ceramic body; dissolving the body and opening up closed air pockets, which allows gas to enter into the glaze.

Ceramic clay and glaze have trace amounts of $CO_3^{2-}$, $SO_4^{2-}$, $NO_3^-$, $Fe_2O_3$, $CaCO_3$, $Pb_3O_4$ and so forth which under high temperatures react and form gasses that create air bubbles. Some of the above-mentioned reactions are listed below:

$CaCO_3 \rightarrow CaO + CO_2\uparrow$ （600 ℃ – 1050 ℃）
$MgCO_3 \rightarrow MgO + CO_2\uparrow$ （400 ℃ – 900 ℃）
$FeCO_3 \rightarrow FeO + CO_2\uparrow$ （>200 ℃）
$2Fe_2O_3 \rightarrow 4FeO + O_2\uparrow$ （1250 ℃ – 1370 ℃）
$Fe_2O_3 + CO \rightarrow 2FeO + CO_2\uparrow$ （1000 ℃ – 1100 ℃）
$CaSO_4 \rightarrow CaO + SO_3\uparrow$ （1250 ℃ – 1370 ℃）
$Na_2SO_4 \rightarrow Na_2O + SO_3\uparrow$ （1200 ℃ – 1370 ℃）

The minerals within the ceramic body and glaze, under high temperatures, discharge water of crystallization that can form air bubbles.

Water particles in the molten bodies, under high temperatures, create air bubbles.

ii. Carbon Formed Air Bubbles

During the firing process, CO undergoes decomposition, splitting to form C and $CO_2$. At the same time, Carbon deposits on the surface of the glaze, oxidize to form $CO_2$, creating air bubbles [87].

iii. Air Bubbles Formed By Technical Influences

During the glazing process, a small portion of air can become trapped within the glaze level; these expand while firing to form air bubbles. During the firing process, combustion products within the kiln furnace can become dislodged into the glaze level and create air bubbles.

iv. Summary

During the firing of ceramics, because of the aforementioned reasons, gasses can enter into the glaze level; some of which seep out of the glaze surface, some of which remain to create air bubbles. The possible gasses within air bubbles are: $O_2$, $H_2$, $H_2O$, $N_2$, CO, $CO_2$, $SO_2$ etc..

While air bubbles are moving throughout the molten glaze, some small bubbles converge with other bubbles to create larger bubbles. Additionally, the coarseness and refinement of the ceramic clay has a direct impact on the size of the bubbles within the glaze. The distribution and size of air bubbles are affected by the following: Firing heat distribution, forging techniques, body and glaze chemical composition, glaze thickness, and so forth.

III. Crystal Phase

a. Crystal Phase Defined

Crystal phase refers to the crystallized appearances of certain colors and shapes across the glaze surface or within the glaze level.

b. Crystal Phase Formation

During the firing process, because of the buoyancy of molten glaze and the stirring of air bubbles, some materials, due to surface tension of the glaze surface, assemble together within the glaze level. These materials are easily assembled near the edges of air bubbles, and some are carried with the air bubbles to the surface of the glaze. When the glaze cools, this matter begins to crystallize within the glaze.

When they crystallize along the surface of the glaze; according to their composition, their density, and differences in surface tension; they present with different appearances, colors, and amounts of mottles (splotches) and striae (lines). These mottles and striae cause unevenness across the glaze surface (see pictures 4-4-8 and 4-4-9). Additionally, some not-fully-melted raw glaze materials remain as particles within the

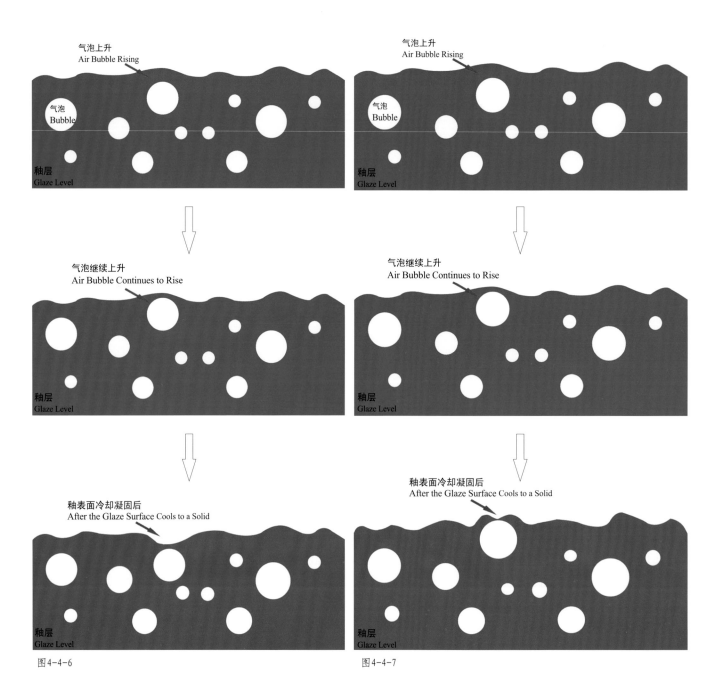

图 4-4-6

图 4-4-7

glaze level.

From a thermodynamics perspective, glaze cooling into a solid does not reach equilibrium and remains in an unstable state, with a tendency to continually have materials separating out and crystallizing. From a kinetics perspective, however, due to the restrained movements of molecules within the cooled solid, crystallization is nearly suspended. Therefore, a vitrified glaze is in a comparatively stable state.

层中的气体。

(2) 气相形成

气相的形成有以下几个原因[86]：

① 陶瓷胎釉本身反应形成气泡

胎体中存在开口气孔和闭口气孔。温度升高时，气体体积膨胀，经开口气孔进入釉层；熔融的釉层将坯体润湿，由于釉对坯体的溶解作用可以打开一些闭口气孔，气体进入釉层。

陶瓷胎釉中含有$CO_3^{2-}$、$SO_4^{2-}$、$NO_3^-$、$Fe_2O_3$、$CaCO_3$、$Pb_3O_4$等，它们在高温下发生化学反应，产生气体，形成气泡。部分反应方程式如下：

$$CaCO_3 \rightarrow CaO + CO_2 \uparrow$$
（600℃-1050℃）

$$MgCO_3 \rightarrow MgO + CO_2 \uparrow$$
（400℃-900℃）

$$FeCO_3 \rightarrow FeO + CO_2 \uparrow$$
（>200℃）

$$2Fe_2O_3 \rightarrow 4FeO + O_2 \uparrow$$
（1250℃-1370℃）

$$Fe_2O_3 + CO \rightarrow 2FeO + CO_2 \uparrow$$
（1000℃-1100℃）

$$CaSO_4 \rightarrow CaO + SO_3 \uparrow$$
（1250℃-1370℃）

$$Na_2SO_4 \rightarrow Na_2O + SO_3 \uparrow$$
（1200℃-1370℃）

陶瓷胎釉中的矿物质在高温下排出结晶水，也会形成气泡被排出。

熔块中的水分在高温下形成气泡。

② 碳素形成气泡

在陶瓷烧制过程中，CO在高温下裂解产生$CO_2$和C，同时C沉积在釉表面，在高温下氧化生成$CO_2$，形成气泡。

③ 工艺影响形成气泡

在施釉时将一部分气体封闭在釉层中，烧制时膨胀形成气泡。在烧制时，窑炉中的燃烧产物会被夹带进入釉层形成气泡。

④ 小结

陶瓷烧制过程中，由上述原因生成的气体进入釉层，部分气体逸出釉面，部分留在釉层中形成气泡。气泡中的气体可以是$O_2$、$H_2$、$H_2O$、$N_2$、CO、$CO_2$、$SO_2$等。

在气泡的运动过程中，一些小气泡会在熔融的釉层中相聚变成大气泡。另外，胎泥的粗与精，对气泡大小的生成有直接关系。烧制火力分布不均匀、工艺参数不同、胎釉的化学组成不同以及釉层的厚度不同等因素，都会影响釉层中气泡的分布和大小。

**3. 晶相**

(1) 晶相概念

晶相是陶瓷釉表面或釉层中析出的有一定形状和颜色的结晶体。

(2) 晶相形成

在陶瓷烧制过程中，熔融的釉料因浮力及气泡的搅动，釉中某些物质在表面张力的作用下在釉层中聚集，也可再形成分相。这些物质易在气泡边缘聚集，部分会随气泡升至釉面。当温度下降时，这些物质便在釉层中析出。

当它们在釉表面析出时，由于它们的成分、密度、表面张力不同，在釉表面形成不同形貌、不同颜色和不同层次的斑块或纹路。这些斑块或纹路会使釉表面凹凸不平（见图4-4-8、图4-4-9）。也有一些釉料中尚未完全熔融的颗粒存在于釉层中。

从热力学的观点来看，冷却凝固了的釉层并未达到平衡，处于介稳状态，仍有继续析出结晶的趋势。但从动力学的观点来看，由于釉层降温固化，分子运动较为困难，析晶几乎停止。所以，冷却凝固了的釉层应处于一个相对稳定的状态。

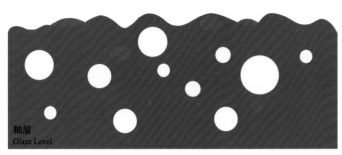

图4-4-8　■析晶或分相 Crystallization or split phase

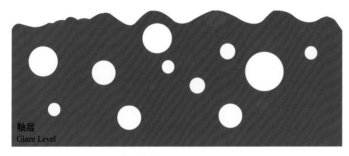

图4-4-9　■析晶或分相 Crystallization or split phase

## 第五节
# 总结

釉层的不同性质是由釉层的成分及其含量决定的。

釉层不同的成分，会形成不同结构的网络，形成釉层不同的性质。如，高温釉主要采用碱金属氧化物及碱土金属氧化物作助熔剂，低温釉主要采用氧化铅为助熔剂。由于使用不同的助熔剂，会形成不同结构的网络，造成了高温釉和低温釉之间性质的差异。

即使釉层成分相同，但各成分含量的差异，也会造成釉层网络结构的差异，形成不同性质的釉层。

此外，工艺参数如烧制温度、烧制气氛等也会对釉层的性质产生影响。

不同性质的釉层会有不同的迹型表现，继而在后天的环境中也会出现不同的风化迹型特征。

一般来说，釉层的风化都是从釉层结构有缺陷（如釉层本身存在的一些裂纹、孔隙）或釉层成分物理化学性质相对较活跃的地方开始（如铁氧化物聚集暴露处），风化侵蚀就从这些地方开始。陶瓷釉层在经过百年甚至千年的风化作用后产生的迹型特征是缓慢量变的结果。现代人为制假做旧，在短时间内以质变、突变的方式制造出的迹型特征，与长期自然风化的迹型特征有本质的区别。

釉层性质的不同是先天性的。釉料配方、工艺参数的差异表现出不同的迹型特征，加上后天环境变化的差异，最终形成釉层迹型风化作用强度快慢的差异和风化强度高低的不同。这样就形成了迹型风化的多样性及复杂性。这是观察、认识和了解陶瓷迹型变化现象以及迹型变化规律的理论基础。

# Section 5:
# Conclusion

The different characteristics of ceramic glaze are determined by the different chemical compositions and respective amounts of individual components.

The different compositions of glaze create different glaze network constructions, and thus different glaze characteristics. High-temperature glazes primarily use alkaline metal oxides and alkaline earth metal oxides as glaze flux, while low-temperature glazes utilize Lead Oxide as flux. The differences in flux compositions create differences among the characteristics of high and low-temperature glazes.

Even though the chemical components of two certain glazes are the same, differences in characteristics can arise from using different amounts of associated components.

Additionally, the technological parameters under which a piece is fired, such as the firing temperature and atmosphere, also have an impact on the characteristics of the glaze.

Different characteristics of glaze present with dissimilar traces, and variances in post-firing environments also affect the trace characteristics of weathering throughout the glaze level.

Usually, weathering first takes place where defects in the glaze level are present, namely crazing and openings; or at points where the physical and chemical compostions of the glaze are more active, i.e. places of exposed Iron Oxides. Corrosion caused by external materials start at these defects. The trace characteristic changes within the glaze level, after hundreds or even thousands of years of weathering, are the result of slow quantitative changes. Modern methods of aging ceramics utilize a qualitative change over a very short period of time and create trace characteristics that are innately differentiated from natural weathering changes.

The differences in glaze characteristics are intrinsic, and these innate differences form unique traces. Additionally, various post-firing environments ultimately bring forth dissimilarities in weathering intensity, and as such, disparities in weathering severity. This brings about multiplicity and complexity in ceramic trace weathering. This is the theoretical foundation for observing, recognizing, and understanding the appearance of trace changes and the laws that govern them.

## 第五章
# 陶瓷釉层变化

**阅读提示：**

　　陶瓷釉层在漫长的岁月里会与周围环境中的各种物质相互作用发生物理变化和化学反应，这个过程叫风化作用。风化作用会导致釉层迹型发生变化。这种变化是在漫长的时间过程中缓慢地、渐进地发生的。可以说，陶瓷釉层迹型变化的本质就是风化的结果。陶瓷釉层的先天性迹型加之风化作用在陶瓷釉层留下的迹型，是判断陶瓷生产年代及生产窑口的客观证据。

**主要论述问题：**

1. 陶瓷釉层在外界环境中会有哪些物理变化和化学反应？
2. 陶瓷釉层的变化与所处环境有什么关系？
3. 陶瓷釉层迹型的风化特征是什么？

## Chapter 5

# Ceramic Glaze Level Changes

**Text Note:**

Over years, the ceramic's glaze level endlessly interacts with all sorts of substances encountered in the surrounding environment; creating physical and chemical changes. This process is called weathering. Weathering brings about the occurrence of trace changes in the glaze level. This is a slow and gradually occurring process of change taking place over a long period of time. We can say ceramic glaze level trace changes are essentially the direct result of weathering. The ceramic glaze level innate traces, as well as the traces acquired through weathering, are the objective evidence we use in determining the period and kiln in which ceramics were made.

# 第一节 陶瓷釉层变化机理

## 一、釉层与水的反应

### （一）高温釉、中温釉与水的反应

由于釉表面的活性，易于吸附水分子，会形成单羟基、双羟基团，单羟基之间会通过氢键键合形成闭合羟基团[87]，反应如下（氢键用虚线表示）：

$\equiv Si—+OH^-\rightarrow \equiv Si—OH$ 　单羟基

$\equiv Si—O^-+H_2O\rightarrow =Si\begin{matrix}OH\\OH\end{matrix}$ 　双羟基团

$\equiv Si—O—H\\\quad\quad\quad\quad\rightarrow$ 闭合羟基团
$\equiv Si—O—H$

通过羟基的氢键再进一步物理吸附水分子，在釉层表面形成水膜。

对高温釉（$Na_2O-K_2O-CaO-MgO-Al_2O_3-SiO_2$系统）、中温釉（$R_2O-SiO_2$系统）来说，釉层中的碱金属离子和碱土金属离子没有参加硅氧网络，均为网络外离子，具有较大的可迁移性。它们会和吸附在釉表面的水分子中的$H^+$进行离子交换，生成碱（ROH）（见反应①），

$\equiv Si—O—R^+ + H^+OH^-\rightarrow$
$\equiv Si—O—H + ROH$ 　　①

接着又有下列反应：

$\equiv Si—O—H + \dfrac{3}{2}H_2O \rightarrow Si(OH)_4$ 　　②

水分子（$H_2O$）也能与硅氧网络直接反应：

$\equiv Si—O—Si\equiv + H_2O \rightarrow$
$2[\equiv Si—O—H]$ 　　③

随着反应③的继续，Si原子周围原有的4个桥氧全部变成—OH，形成$Si(OH)_4$，这是水对硅氧网络的直接破坏[88]。

以上反应的产物$Si(OH)_4$是一种极性分子，它能使周围的水分子极化，然后吸附在自己周围，成为$Si(OH)_4\cdot nH_2O$，通常称为硅酸凝胶[89]。这种硅酸凝胶除了有一部分溶于水之外，大部分都会附着在釉表面，形成一层薄膜。它具有较强的抗水和抗酸能力。

反应①中，由于$R^+$比$H^+$要大，它们交换后，$H^+$代替$R^+$，使釉层产生空位，结构变疏松，稳定性下降，所以易受到风化侵蚀。

### （二）低温釉、中温釉与水的反应

低温釉（$PbO-Al_2O_3-SiO_2$或$PbO-SiO_2$系统）和中温釉（$R_2O-PbO-SiO_2$系统）与水的反应机理与上述的高温釉、中温釉基本相同。

中温釉（$R_2O-PbO-SiO_2$系统）中的$R^+$会和吸附在釉表面的水分子中的$H^+$进行离子交换，生成ROH（见反应①），进而产生硅酸凝胶（见反应②）。

中温釉（$R_2O-PbO-SiO_2$系统）中R—O的键强比Pb—O要小的多（键强值KJ/mol：Na—O：84，K—O：54，Pb—O：151）[90]，所以$R^+$的迁移更容易，它们迁出釉层后会留下一些空位，釉层中的$Pb^{2+}$便会逐渐占据这些空位[91]，慢慢迁移到釉表面。即使$R_2O$含量很低，也能够加快$Pb^{2+}$的迁出。所以含有碱金属氧化物（$R_2O$）的铅釉更容易受到风化侵蚀。

此外，低温釉和中温釉中的$Pb^{2+}$也会与水中的$H^+$（或水合氢离子$H_3O^+$）进行离子交换，$Pb^{2+}$迁出，生成难溶$Pb(OH)_2$（见反应④）：

$\equiv Si—O—Pb—O—Si\equiv + 2H_2O \rightarrow$
$2\equiv Si—O—H + Pb(OH)_2$ 　　④

$Pb(OH)_2$通常是白色物质，易

# Section 1:
# Mechanics of Glaze Level Change

## 1. Glaze Level Reactions with Water

### A. High and Medium-Temperature Glaze Reactions with Water

The active nature of the glaze surface and its high absorption allows for the formation of single and double hydroxyls, as well as the hydrogen bonding between single hydroxyls forming closed hydroxyl groups [87]. The reactions are shown below (Hydrogen bonding is represented with a dotted line):

$\equiv$Si—+OH$^-$→$\equiv$Si—OH        Single Hydroxyl

$\equiv$Si—O$^-$+H$_2$O→$\equiv$Si$\underset{\diagdown \text{OH}}{\overset{\diagup \text{OH}}{}}$        Double Hydroxyls

$$\equiv\text{Si—O—H} \atop \equiv\text{Si—O—H} \quad \rightarrow \quad \begin{array}{c} \text{H} \quad \text{H} \\ \text{O} \quad \text{O} \\ | \quad | \\ \text{Si} \quad \text{Si} \\ /|\diagdown/|\diagdown \\ \text{O} \end{array} \quad \text{Hydroxyl Group}$$

The Hydrogen bonds of these hydroxyls physically attract water molecules, forming a film of water over the surface of the glaze.

As the alkaline metal and alkaline earth metal ions in high-temperature (Na$_2$O-K$_2$O-CaO-MgO-Al$_2$O$_3$-SiO$_2$ system) and medium-temperature (R$_2$O-SiO$_2$ system) glazes do not take part in the silica network, they retain a relatively high ability to migrate throughout the glaze level. This allows them to partake in ion exchange with the H$^+$ ion of the water that has been absorbed to the glaze surface, creating alkali (ROH) (see reaction ①).

$\equiv$Si—O—R$^+$ + H$^+$OH$^-$ → $\equiv$Si—O—H + ROH     ①

Additionally there are the below reactions:

$\equiv$Si—O—H + $\dfrac{3}{2}$H$_2$O → Si(OH)$_4$     ②

Water particles can also react directly with the silica network:

$\equiv$Si—O—Si$\equiv$ + H$_2$O → 2[$\equiv$Si—O—H]     ③

The continuation of reaction ③ is the four bridging Oxygen surrounding the Si atom becoming —OH particles, forming Si(OH)$_4$. This is the direct breakdown effect water has on the silica network [88].

The Si(OH)$_4$ created in the above reaction is a polarized molecule that can also polarize the surrounding water molecules, attracting them to itself creating Si(OH)$_4$•nH$_2$O, often called silicic acid gel [89]. Aside from a small portion that dissolves in water, this silicic acid gel can absorb to the glaze surface, forming a thin layer or film. This film, in turn, has a relatively high resistance to water and other acids.

In reaction ①, the location left by the R$^+$ ion, being larger than the H$^+$ ion with which they change places, leaves voids in the glaze level. This in turn, loosens the structure, decreases the stability, and increases the susceptibility to corrosion.

### B. Low and Medium-Temperature Glaze Reactions with Water

The reaction mechanics between water and low-temperature (PbO-Al$_2$O$_3$-SiO$_2$ or PbO-SiO$_2$ system) and

溶于酸，微溶于弱碱，而大量溶于强碱介质中。

由于$Pb^{2+}$比$H^+$大，它们交换后，$H^+$代替$Pb^{2+}$，使釉层产生空位，结构变疏松，稳定性下降，所以更易受到风化侵蚀。

## 二、釉层与碱的反应

碱的来源有两个方面，一是来自反应①，水与釉表面作用生成的，它取决于$H^+$和$R^+$交换的速度，二是外部环境提供的。

碱（ROH）可溶解硅酸凝胶$Si(OH)_4 \cdot nH_2O$（见反应⑤），使釉表面失去保护，风化得以继续进行。

$$Si(OH)_4 + ROH \rightarrow [Si(OH)_3]^- R^+ + H_2O \quad ⑤$$

ROH也可以直接与硅氧网络反应，使Si—O键断裂，网络解体，产生$\equiv Si-O^-$粒子群，使$SiO_2$溶于碱液中，因而破坏釉表面（见反应⑥）：

$$\equiv Si-O-Si \equiv + OH^- \rightarrow \equiv Si-OH + \equiv Si-O^- \quad ⑥$$

而断裂一端的$\equiv Si-O^-$又能与$H_2O$反应：

$$\equiv Si-O^- + H_2O \rightarrow \equiv Si-OH + OH^- \quad ⑦$$

反应⑦产生的碱可使反应⑥继续发生，从而溶解$SiO_2$。

### （一）高温釉、中温釉与碱的反应

对于高温釉（$Na_2O-K_2O-CaO-MgO-Al_2O_3-SiO_2$系统）而言，随着$R^+$（R在此指Na、K）的不断迁出，$Ca^{2+}$的含量相对增多，$Ca^{2+}$会对$R^+$产生"抑制效应"，使得$H^+$和$R^+$的离子交换变得缓慢，碱（ROH）的生成速度变慢（反应①速度变慢），从而反应⑥也变得缓慢。这样对釉层的侵蚀就变得非常缓慢，甚至几乎停止了。但如果外部环境是碱性的，风化作用会继续进行。

对于中温釉（$R_2O-SiO_2$系统）而言，由于没有$Ca^{2+}$的"抑制效应"，$R^+$会不断迁出，一直到釉层被完全侵蚀。

由于碱（ROH）的作用，硅氧网络遭到破坏后（见反应⑥），以$\equiv Si-O^-$粒子群的形式溶解下来，和釉表面游离的$R^+$、$R^{2+}$等阳离子结合成硅酸盐。这些风化产物不断堆积，在釉表面形成一个壳层。当这一壳层达到一定厚度时，会令釉表面失去光泽，严重时呈不透明状。此壳层是疏松多孔的，吸水性也很强，它提供了一个有高浓度的碱性水溶液的环境，使釉表面无法形成保护膜，原有的硅酸凝胶薄膜也会遭到破坏。因此，这种壳层的形成会大大加剧外界对釉层的侵蚀[92]。

### （二）低温釉、中温釉与碱的反应

碱液同样会使低温釉（$PbO-Al_2O_3-SiO_2$或$PbO-SiO_2$系统）和中温釉（$R_2O-PbO-SiO_2$系统）的网络结构遭到破坏，除了生成硅酸盐外，还会与$Pb^{2+}$生成难溶的$Pb(OH)_2$（见反应⑧），附着在釉表面上。$Pb(OH)_2$遇到环境中的酸性或强碱溶液时，便会溶解。

$$Pb^{2+} + 2OH^- \rightarrow Pb(OH)_2 \quad ⑧$$

## 三、釉层与酸的反应

### （一）高温釉、中温釉与酸的反应

釉层与酸的反应跟其与水的反应相似，除氢氟酸外，一般的酸并不直接与釉发生反应，而是通过水起作用。

氢氟酸对玻璃的作用是直接破坏硅氧网络（见反应⑨）。

$$\equiv Si-O-Si \equiv + H^+F^- \rightarrow \equiv Si-O-H + \equiv Si-F \quad ⑨$$

对高温釉（$Na_2O-K_2O-CaO-MgO-Al_2O_3-SiO_2$系统）和中温釉（$R_2O-SiO_2$系统）来说，酸溶液中的$H^+$浓度比水中的$H^+$浓度要大，所以$H^+$与$R^+$的离子交换速度在酸中比在水中快，即反应①有较快的速度，生成的碱（ROH）会侵蚀釉表面。如果$R_2O$含量低或有$Ca^{2+}$的"抑制效应"，反应①就会减缓速度或停止。

$Si(OH)_4$不溶于酸，在酸性溶液中溶解度减小，即减慢了⑤的反应速度，从而能够减缓釉表面的侵蚀。

在高碱釉中，$R_2O$含量较高时，$R^+$的浓度较高，反应①是主要的，有碱不断生成，侵蚀釉表面；在高硅釉中，$SiO_2$含量较高时，反应⑤的速度会变得缓慢，也就是说硅酸凝胶被溶解的速度减慢，风化得以减缓。所以，高碱釉的耐酸性小于耐水性，而高硅釉的耐酸性大于耐水性。

medium-temperature ($R_2O$-PbO-$SiO_2$ system) glazes is essentially the same as the aforementioned high and medium-temperature glazes.

The $R^+$ ions of medium-temperature glazes swap $H^+$ ions with the water molecules absorbed by the glaze surface, creating ROH (see reaction ①) and form silicic acid gel (see reaction ②).

The bonding strength between R—O of medium-temperature glaze is much lower than that of Pb—O (bond strength Kj/mol: Na—O: 84, K—O: 54, and Pb—O: 151) [90]. Therefore it is much easier for $R^+$ ions to migrate; leaving voids in the glaze level as they migrate out. In which case, the $Pb^{2+}$ ions in the glaze level gradually occupies those voids [91], and migrates to the glaze surface. Even a trace amount of $R_2O$ can accelerate the migration of $Pb^{2+}$ ions, therefore Lead glazes containing alkaline oxides easily corrode under weathering.

Additionally, the $Pb^{2+}$ ions in low and medium-temperature glazes can enter into ion exchange with surface $H^+$ ions (or hydronium ion $H_3O^+$), in which $Pb^{2+}$ ions migrate out and create $Pb(OH)_2$ (see reaction ④):

$$\equiv Si-O-Pb-O-Si\equiv + 2H_2O \rightarrow 2\equiv Si-O-H + Pb(OH)_2 \quad ④$$

$Pb(OH)_2$ often presents as a white colored material. It is acid soluble, slightly soluble in weak bases, and highly soluble in strong bases.

The location left by $Pb^{2+}$ ions, being larger than the $H^+$ ions with which they change places, leaves voids in the glaze level. This in turn loosens the structure, decreases the stability, and increases the susceptibility to corrosion.

## 2. Glaze Level Reactions with Alkali

The introduction of alkali occurs in two different ways: The first being the aforementioned reaction ① water-glaze reaction, depending on the speed of $R^+$ and $H^+$ ion exchange. The second is the introduction via external environment.

Alkali (ROH) can dissolve silicic acid gel (see reaction ⑤); this destroys the glaze surface protection, and the weathering process continues.

$$Si(OH)_4 + ROH \rightarrow [Si(OH)_3O]^- R^+ + H_2O \quad ⑤$$

ROH can also react directly with the silica network, breaking down the Si—O bonds, creating Si—$O^-$ grains, dissolving the $SiO_2$ in alkali solution, and breaking down the glaze surface (see reaction ⑥):

$$\equiv Si-O-Si\equiv + OH^- \rightarrow \equiv Si-OH + \equiv Si-O^- \quad ⑥$$

The other side of the $\equiv Si-O^-$ split is able to react with water:

$$\equiv Si-O^- + H_2O \rightarrow \equiv Si-OH + OH^- \quad ⑦$$

The alkali formed in reaction ⑦ can facilitate the continuation of reaction 6, and break down the glaze surface.

## A. High and Medium-Temperature Glaze Reactions with Alkali

In reference to the continuous migrating out of $R^+$ ions (where R represents Na and K) in high-temperature glazes ($Na_2O$-$K_2O$-CaO-MgO-$Al_2O_3$-$SiO_2$ system), $Ca^{2+}$ ions in relatively high numbers can act as reaction inhibitors; slowing the $H^+$ and $R^+$ ion exchange, slowing the creation of alkali (ROH) (reaction ① slows), and slowing the occurrence of reaction ⑥. This drastically reduces the speed at which the glaze surface corrodes, almost bringing it to a stop. However, if the external environment has an alkaline nature, weathering can proceed.

In regards to medium-temperature glazes ($R_2O$-$SiO_2$ system), the lack of $Ca^{2+}$ ions and their inhibiting nature allows for the continuous $R^+$ ion migration, without interruption, until the glaze level is completely corroded.

With the introduction of alkali (ROH), after the silica network encounters breakdown (see reaction ⑥), $\equiv Si-O^-$ grains dissolve down together with the free $R^+$ and $R^{2+}$ ions on the surface and join to create silicate salts. This weathering causes continuous sedimentation, forming a sort of shell over the glaze level. As this shell reaches a certain thickness, it depletes the sheen of the glaze surface and decreases its translucency.

### （二）低温釉、中温釉与酸的反应

对于低温釉（$PbO-Al_2O_3-SiO_2$或$PbO-SiO_2$系统）和中温釉（$R_2O-PbO-SiO_2$系统）而言，机理与上述的高温釉、中温釉相似。但生成的碱中还有难溶的$Pb(OH)_2$，$Pb(OH)_2$溶于酸。随着$Pb^{2+}$从釉层中迁出，还会与$CO_3^{2-}$、$SO_4^{2-}$等酸根反应，生成难溶盐（见反应⑩、⑪）。

所以，酸性环境能够加快低温釉（$PbO-Al_2O_3-SiO_2$或$PbO-SiO_2$系统）和中温釉（$R_2O-PbO-SiO_2$系统）陶瓷的风化。

$$Pb^{2+} + CO_3^{2-} \rightarrow PbCO_3 \downarrow \quad ⑩$$

$$Pb^{2+} + SO_4^{2-} \rightarrow PbSO_4 \downarrow \quad ⑪$$

## 四、釉层与$CO_2$的反应

### （一）高温釉、中温釉与$CO_2$的反应

$CO_2$会与釉表面的碱反应生成碳酸盐（$Na_2CO_3$或$NaHCO_3$）（见反应⑫、⑬）：

$$NaOH + CO_2 \rightarrow NaHCO_3 \quad ⑫$$

$$2NaOH + CO_2 \rightarrow Na_2CO_3 + H_2O \quad ⑬$$

它们在釉表面会进一步吸附水分子，形成随机分布的白色雾状斑点，属于可溶盐[93]。当水分蒸发后，这些盐便析出，它能稳定存在于空气中。由于$CO_2$与碱反应生成盐，降低了釉表面的$R^+$和$OH^-$的浓度，这更有利于反应①中的离子交换。而釉表面的碱被中和成盐，减缓了碱对釉层的直接侵蚀，有利于硅酸凝胶的形成与保持。

但是，当釉表面吸附水分子时，处于釉表面的碳酸盐结晶颗粒（$Na_2CO_3$或$NaHCO_3$）又会水解生成碱（ROH），ROH溶解硅酸凝胶（见反应⑤），侵蚀釉表面。

$CO_2$还会与析出到釉表面的游离$Ca^{2+}$等作用形成碳酸盐晶体，属于难溶盐（见反应⑭）：

$$Ca^{2+} + CO_3^{2-} \rightarrow CaCO_3 \downarrow \quad ⑭$$

### （二）低温釉、中温釉与$CO_2$的反应

$CO_2$会与釉表面的$Pb(OH)_2$及$Pb^{2+}$发生反应，生成Pb的碳酸盐（见反应⑮、⑯）：

$$Pb(OH)_2 + CO_2 \rightarrow PbCO_3 \downarrow + H_2O \quad ⑮$$

$$Pb^{2+} + CO_3^{2-} \rightarrow PbCO_3 \downarrow \quad ⑯$$

## 五、釉层与盐溶液的反应

釉层中迁出的金属离子会与盐溶液的酸根（如$CO_3^{2-}$，$SO_4^{2-}$，$SiO_3^{2-}$，$PO_4^{3-}$）发生反应，生成可溶盐或难溶盐。部分反应如下：

$$Na^+ + CO_3^{2-} \rightarrow Na_2CO_3 \quad ⑰$$

$$Ca^{2+} + CO_3^{2-} \rightarrow CaCO_3 \downarrow \quad ⑱$$

$$Pb^{2+} + SO_4^{2-} \rightarrow PbSO_4 \downarrow \quad ⑲$$

$$Ca^{2+} + SO_4^{2-} \rightarrow CaSO_4 \quad ⑳$$

$$Pb^{2+} + CO_3^{2-} \rightarrow PbCO_3 \downarrow \quad ㉑$$

在干湿交替的环境中，可溶盐如$Na_2CO_3$、$NaHCO_3$可能流失，也可能渗入釉裂纹中，当水分蒸发后便结晶析出。难溶盐会附着于釉表面或沉积于釉裂纹中。

## 六、釉层风化产物

陶瓷所存在的后天环境是复杂多样的，很多时候风化产物并不仅仅是上述所有反应产物的简单叠加。

以下关于我国古代历代低温釉风化产物的研究充分地说明了这一点。

朱铁权等人的研究表明[94]：

汉代绿釉陶表面的腐蚀物主要为碳酸铅（$PbCO_3$），以白铅矿的物相结构存在；

宋代绿釉陶表面的腐蚀物主要为磷酸铅钙[$Pb_{10-x}Ca_x(PO_4)(OH)_2$（x<2.7）]，锈蚀物中黄色是缘自其中一定量氧化铁或者铁的化合物存在；

唐代唐三彩表面的腐蚀物中有磷酸铅钙与白铅矿两种物相存在，锈蚀物表面的棕褐色可能是少量铁与炭黑共同致色的结果；

不同种类的铅釉陶表面腐蚀物中存在磷酸铅与磷酸铅钙，它们是釉中的铅离子与埋藏环境中磷酸根、碳酸根发生化学反应的结果。

各种器物的制作工艺以及埋藏环境的不同，其腐蚀物的物相组成有所差异。

## 七、小结

釉层的风化过程是一个漫长、复杂的物理变化和化学反应的综合过程，上述的各种化学反应并非是

This shell is loosely porous and highly water absorbent; creating a strong alkali solvent liquid covering, disallowing the glaze to form its protective film, and destroying the silicic acid gel film originally present. Because of this shell, the corrosion to the glaze surface by the external environment is drastically increased [92].

## B. Low and Medium-Temperature Glaze Reactions with Alkali

Alkali liquids have a similar corrosive effect on low-temperature ($PbO-Al_2O_3-SiO_2$ or $PbO-SiO_2$ system) and medium-temperature ($R_2O-PbO-SiO_2$ system) glazes. Aside from creating silicate salts, it can react with $Pb^{2+}$ ions to form partially soluble $Pb(OH)_2$ (see reaction ⑧). Absorbed by the glaze surface, this $Pb(OH)_2$ can dissolve if met with an acidic or a high alkali solvent environment.

$$Pb^{2+} + 2OH^- \rightarrow Pb(OH)_2 \qquad ⑧$$

## 3. Glaze Level Reactions with Acid

### A. High and Medium-Temperature Glaze Reactions with Acid

Glaze level and acid reaction are very similar to that of reactions with water. Aside from hydrofluoric acid, normal acids don't directly react with the glaze surface, but rely on water to react.

The effect of hydrofluoric acid on glass is direct destruction to the silica network (see reaction ⑨).

$$\equiv Si-O-Si\equiv + H^+F^- \rightarrow \equiv Si-O-H + \equiv Si-F \qquad ⑨$$

For high-temperature ($Na_2O-K_2O-CaO-MgO-Al_2O_3-SiO_2$ system) and medium-temperature ($R_2O-SiO_2$ system) glazes, the amount of $H^+$ ions in acid solutions is much higher than that of water, therefore the $R^+$ and $H^+$ ion exchange rate is much higher than that of water; accelerating reaction ① and creating glaze-corrosive alkali (ROH). If the amount of $R_2O$ is low or the $Ca^{2+}$ ion's inhibiting effect is present, reaction ① slows or even stops.

$Si(OH)_4$ is not acid soluble, therefore its solubility in acid solution is very low, which can slow reaction ⑤ and slow glaze surface corrosion.

In high alkaline glazes, $R_2O$ is relatively high and $R^+$ ions are abundant. Reaction ① is predominant, and the continuous formation of alkali corrodes the glaze surface. In high silicone glazes, $SiO_2$ is relatively high. This slows reaction ⑤; which is to say the dissolving of the silicic acid gel slows, thus slowing the weathering process. Therefore, high alkaline glazes are more water-resistant than acid-resistant, and high silicone glazes are more acid-resistant than water-resistant.

### B. Low and Medium-Temperature Glaze Reactions with Acid

The reaction mechanics of low ($PbO-Al_2O_3-SiO_2$ or $PbO-SiO_2$ systems) and medium-temperature ($R_2O-PbO-SiO_2$ system) glazes with acid is similar to the previously mentioned reactions with high and medium-temperature glazes. However the formed alkali also present with $Pb(OH)_2$, which is acid soluble. In addition, as $Pb^{2+}$ ions move out from the glaze; they can react with acid radicals such as $CO_3^{2-}$ and $SO_4^{2-}$, creating partially soluble salts (see reactions ⑩ and ⑪).

Therefore, acidic environments can speed up the breakdown of low and medium-temperature glazes ($PbO-Al_2O_3-SiO_2$, $PbO-SiO_2$ or $R_2O-PbO-SiO_2$ systems).

$$Pb^{2+} + CO_3^{2-} \rightarrow PbCO_3\downarrow \qquad ⑩$$

$$Pb^{2+} + SO_4^{2-} \rightarrow PbSO_4\downarrow \qquad ⑪$$

## 4. Glaze Level Reactions with Carbon Dioxide

### A. High and Medium-Temperature Glaze Reactions with $CO_2$

Carbon Dioxide can react with the alkali on the glaze surface creating carbonates ($Na_2CO_3$ or $NaHCO_3$) (see reactions ⑫ and ⑬).

一个单一过程，可能会有几种反应同时进行：如当$H^+$与$R^+$产生交换的同时，也有ROH与釉层的反应，ROH还可能同时与$CO_2$发生反应；同时也有物理变化发生，如盐溶液中的晶体析出，温度变化使釉层产生裂纹等。

由于陶瓷所存在的后天环境是复杂多样的，风化的产物也多种多样，所以，陶瓷迹型也呈现出多样性。

$$NaOH + CO_2 \rightarrow NaHCO_3 \qquad ⑫$$

$$2NaOH + CO_2 \rightarrow Na_2CO_3 + H_2O \qquad ⑬$$

Carbonates act to absorb water on the glaze surface forming randomly distributed, foggy-white blotches, which are soluble salts that crystallize as the water evaporates and stabilize in open air [93]. As the Carbon Dioxide and alkali form salts, the amount of $R^+$ and $OH^-$ ions decrease, fostering the ion exchange in reaction ①. The neutralization of alkali into salt on the glaze surface slows its ability to directly corrode the glaze surface, and promotes the formation of the protective silicic acid gel.

However, once the glaze surface absorbs water, surface level crystallized salts ($Na_2CO_3$ or $NaHCO_3$) again dissolve into alkali (ROH); which in turn dissolves the silicic acid gel (see reaction ⑤) and corrodes the glaze surface.

Carbon Dioxide can also join with surface level free ions such as $Ca^{2+}$ and form carbonate crystals which are partially soluble salts (see reaction ⑭).

$$Ca^{2+} + CO_3^{2-} \rightarrow CaCO_3\downarrow \qquad ⑭$$

## B. Low and Medium-Temperature Glaze Reactions with $CO_2$

Carbon Dioxide can react with surface level $Pb(OH)_2$ and $Pb^{2+}$ ions to form Lead Carbonates (see reaction ⑮ and ⑯):

$$Pb(OH)_2 + CO_2 \rightarrow PbCO_3\downarrow + H_2O \qquad ⑮$$

$$Pb^{2+} + CO_3^{2-} \rightarrow PbCO_3\downarrow \qquad ⑯$$

## 5. Glaze Level Reactions with Salt Solutions

The migrating metal ions on the glaze surface can react with the acid radicals (i.e. $CO_3^{2-}$, $SO_4^{2-}$, $SiO_3^{2-}$, $PO_4^{3-}$) in salt solutions to form soluble and partially soluble salts. Sample reactions:

$$Na^+ + CO_3^{2-} \rightarrow Na_2CO_3 \qquad ⑰$$

$$Ca^{2+} + CO_3^{2-} \rightarrow CaCO_3\downarrow \qquad ⑱$$

$$Pb^{2+} + SO_4^{2-} \rightarrow PbSO_4\downarrow \qquad ⑲$$

$$Ca^{2+} + SO_4^{2-} \rightarrow CaSO_4\downarrow \qquad ⑳$$

$$Pb^{2+} + CO_3^{2-} \rightarrow PbCO_3\downarrow \qquad ㉑$$

In environments of alternating moisture, soluble salts such as $Na_2CO_3$ and $NaHCO_3$ can dissolve away. It is also possible that they permeate into glaze cracks and crystallize when the water evaporates. Partially soluble salts can become absorbed to the glaze surface or deposit into glaze cracks.

## 6. Sedentary Byproducts

The environments in which ceramics pass time are complex and diverse. Often the weathering byproducts are not confined to the simple super impositions of the aforementioned reactions.

This is clearly illustrated by the research done by Zhu et al on Ancient Chinese low-temperature glaze ceramics as presented below [94]:

The Han Dynasty Green Glaze Ceramics' primary sedentary byproducts consist of Lead Carbonate ($PbCO_3$), presenting as white Lead.

The Song Dynasty Green Glaze Ceramics' primary sedentary byproducts consist of Lead Calcium Phosphate [$Pb_{10-x}Ca_x(PO_4)(OH)_2(x<2.7)$]; yellow rust spots present because of the presence of iron oxide or other iron compounds;

The Tang Dynasty Three-Color Ceramics glaze level sedentary byproducts include Lead Calcium Phosphate and white Lead, the brown rust coloration possibly comes from the mixture of iron and charcoal.

The Lead Carbonate and Lead Calcium Phosphate discovered in these lead glazes was formed by the combination of Lead ions with Carbonate and Phosphate radicals present in the surrounding environment.

Due to the differences in the manufacturing and storage

of each piece, there is a vast variation in the sedentary byproducts that present.

## 7. Summary

Glaze level weathering is the accumulation of a lengthy, complicated physical and chemical change process. The above mentioned chemical changes are not entirely independent processes, but rather a series of simultaneously occurring events: For example, as the $R^+$ and $H^+$ ions exchange, the ROH and glaze level reaction is taking place; ROH could also be reacting with Carbon Dioxide. Meanwhile, physical changes can be occurring: For example the crystallization of salt solutions or temperature changes creating glaze cracks.

Due to the multi-faceted and complex characteristics of post-firing environments, the materialization of weathering is not limited to the aforementioned reactions, but can contain other byproducts. Therefore, ceramic traces also present with multiple variations.

# 第二节 陶瓷釉层变化与环境关系

## 一、地上环境

### （一）地上环境概念

地上环境是指陶瓷所处的自然大气情况。

### （二）地上环境组成

大气的主要成分是氮$N_2$（78.08%），氧$O_2$（20.95%）、氩$Ar$（0.934%）和$CO_2$（0.034%）。大气中也含有水，但水的含量是一个可变化的数值，在不同的时间、不同的地点及不同的气候条件下，水的含量是不一样的，其数值一般在1%-3%的范围内发生变化。此外，大气中还有很多痕量组分，如$H_2$、$CH_4$、CO、$SO_2$、$NH_3$、$N_2O$、$NO_2$、$O_3$[95]。地上环境中还有细小灰尘，它们会被吸附或沉积在釉表面上。

### （三）地上环境的风化

陶瓷釉表面在地上环境中的风化实质是水、$CO_2$、$SO_2$等在釉表面作用的总和。釉表面灰尘积聚处易吸附大气中的水分，有利于风化的进行。

由第一节可知，高温釉、中温釉表面会吸附空气中的水分子，在釉表面形成水膜；通过$H^+$和$R^+$离子交换等一些反应，釉表面会产生硅酸凝胶薄膜（见反应①②③）。如果釉层的$R_2O$含量少，这种薄膜的形成及发展非常缓慢；如果$R_2O$含量较高，迁出的$R^+$则形成ROH（见反应①），ROH会溶解附在釉表面的硅酸凝胶薄膜（见反应⑤），接着直接与釉表面的硅氧网络反应，使网络解体（见反应⑥），破坏釉表面。

风化后，通常有雾状或颗粒状白斑，自然分布在釉表面或深入釉层中。

中、低温釉也有类似的情况，析出物还可为银屑状，是$Pb(OH)_2$及Pb的碳酸盐等。

## 二、地下环境

### （一）地下环境概念

地下环境是指陶瓷所处的土壤情况。

### （二）地下环境组成

土壤是由固体、液体和气体三相共同组成的多相体系。固相是指矿物质和有机质，矿物质约占90%以上，有机质占1%-10%。液相是指土壤中的水分及其水溶物。气相是指充斥在土壤无数孔隙中的空气。

不同于在大气中，土壤中的水是具有一定酸碱性的。我国土壤的pH值大多在4.5-8.5范围内，并有由南向北pH值递增的规律性：如华南，西南地区大多为pH值4.5-5.5，有少数低至3.6-3.8；华中、华东地区为5.5-6.5；长江以北的土壤多为中性或碱性，如华北、西北的土壤大多含$CaCO_3$，pH值一般在7.5-8.5，少数强碱性土壤的pH值高达10.5[96]。

### （三）地下环境的风化

#### 1. 液相产生的风化作用

在土壤中由液相产生的风化作用是以水溶液的形式与釉表面进行的。风化作用可以是$H^+$与碱金属离子交换，其结果是釉层产生空位，结构变疏松（见反应①），也可以是碱直接破坏硅氧网络（见反应②），或者是生成难溶盐沉积在釉表面，或生成可溶盐流失。

水溶液中的可溶盐可渗入釉裂纹、孔隙等釉层的疏松结构中，水分蒸发后，可溶盐结晶析出。在干湿交替的过程中，溶解、结晶反复进行，产生的应力会使釉表面开裂，甚至出现剥釉。

土壤中的及釉表面的阴阳离子（$Ca^{2+}$、$Mg^{2+}$、$Fe^{3+}$、$Mn^{2+}$、$Pb^{4+}$、$CO_3^{2-}$、$SO_4^{2-}$、$OH^-$、$SiO_3^{2-}$、$PO_4^{3-}$）

# Section 2:
# Relationship Between Environment and Glaze Level Change

## 1. Above Ground Environment

### A. Above Ground Environment Defined

Above Ground Environment indicates a positioning situation in which a ceramic piece is exposed to open air.

### B. Above Ground Environmental Components

Atmosphere (or open air) consists of primarily $N_2$ (78.08%), $O_2$ (20.95%), Ar (0.934%), and $CO_2$ (0.034%). Open air also includes a certain amount of water, but to varying degrees. Due to humidity changing over time and different weather conditions, water usually compiles 1%-3% of the atmosphere. In addition, open air often includes trace amounts of the following compounds: $H_2$, $CH_4$, CO, $SO_2$, $NH_3$, $N_2O$, $NO_2$, $O_3$ [95]. The above ground environment also includes very fine dust particles, which can be absorbed by or deposited onto the glaze surface.

### C. Above Ground Environment Weathering

Above ground weathering of ceramic glaze is the accumulation of the interaction between the glaze level and water, $CO_2$, $SO_2$ and so forth. Glaze areas that have dust deposits tend to absorb water; creating a starting point for weathering.

From Section 1, we know that high and medium-temperature glaze surfaces absorb water, creating a sort of wet film across the surface. Through $H^+$ and $R^+$ ion exchange, the surface can present with a thin silicic acid gel covering (see reactions ①②③). If the amount of $R_2O$ is relatively low, the formation of this film is extremely slow. If the amount of $R_2O$ is relatively high, free $R^+$ ions separate out to form ROH (see reaction ①), which in turn dissolves the silicic acid gel film (see reaction ⑤). The result is direct contact between ROH and the silica network, and the breaking down and erosion of the glaze surface (see reaction ⑥).

After weathering, it is common to see foggy-white and grainy-white mottles naturally occurring across the glaze surface, and entering into the glaze level.

Low and medium-temperature glazes are similar, crystallized materials being $Pb(OH)_2$, Lead Carbonate, and so forth.

## 2. Underground Environment

### A. Underground Environment Defined

Underground Environment indicates a positioning situation in which a ceramic piece is under the surface of the earth.

### B. Underground Environmental Components

Soil is a multi-phase body that consists of conglomeration of gas, liquid, and solid phases. Solids consist of primarily +90% minerals and 1%-10% organics. Liquids are a combination of water and hydrotropes. Gasses are comprised of a combination of any air components that exist in the numerous cavities within the soil.

Unlike open-air moisture, the water within soil has certain acidic or basic properties. Chinese soil for example primarily ranges from 4.5-8.5 pH, with the pH generally increasing as you move toward northern latitudes: Southern and Southwestern China soil primarily ranges between 4.5-5.5 pH, with a small portion between 3.6-3.8 pH; Central

会发生反应，形成不同颜色的难溶盐沉积物。难溶盐在沉积过程中还可胶结土壤颗粒，在釉表面形成坚硬的垢层。

水溶液侵蚀釉表面，从而使釉层变成疏松的多孔层，这时，如果有土壤细小颗粒进入釉层中，待水分蒸发时，便会形成土沁，颜色与周围土壤接近。

土壤中各种矿物质颗粒并非均匀分布，而由于土壤颗粒表面的吸附力和微细孔隙的毛细管作用，水分被保持在土壤中不流动，这使土壤中各处的水溶液成分及干湿程度有差异。所以，釉表面各处被风化侵蚀的方式和程度就有所不同。

### 2. 气相产生的风化作用

土壤中的空气组成与大气基本相似，主要成分都是$N_2$、$O_2$、$CO_2$。其差异是：

（1）土壤空气存在于相互隔离的土壤孔隙中，是一个不连续的体系。

（2）在$O_2$和$CO_2$含量上有很大差别，土壤空气中的$CO_2$含量比大气中高得多，大气中$CO_2$含量仅为0.02%—0.03%，而土壤空气中$CO_2$含量一般在0.15%—0.65%，甚至高达5%，它主要是由生物呼吸作用和有机物分解产生；土壤中的氧气含量低于大气，而水蒸气的含量比大气中高得多；土壤空气中还含有少量还原性气体，如$CH_4$、$H_2S$、$H_2$、$NH_3$等[97]。

土壤中的$CO_2$和水的含量比大气中要高得多，存在于土壤中的陶瓷风化强度比大气中的要高很多。

## （四）小结

地下环境的复杂性及不均匀性，造成陶瓷釉表面风化的随机性：有的地方风化严重，有的地方保存相对完好。

潮湿的土壤比干燥的土壤腐蚀性强，干湿交替的土壤比湿度恒定的土壤腐蚀性强，酸性或碱性土壤比中性土壤腐蚀性强。与地上风化形貌最大的差异是，由于有难溶盐及土壤小颗粒的存在，地下环境的风化产物可以是黄色、褐色或其他颜色。

如果在密封较好、较干燥的地下环境中，瓷器的风化强度要比直接接触土壤的瓷器低。

## 三、水中环境

### （一）水中环境概念

水中环境是指陶瓷所处的水下情况。

### （二）水中环境风化

当陶瓷处在水中时，釉层中的网络外的金属离子（如$Na^+$、$K^+$、$Pb^{2+}$等）与$H^+$交换，生成碱。一般情况下，碱会被水稀释，因而不能进一步腐蚀釉表面。但如果$R_2O$含量较高，则$R^+$与$H^+$的交换会继续进行，使釉表面变得疏松多孔，最终导致釉层脱落。同样的，$Pb^{2+}$在水中也会不断地迁出，长期如此，釉层会产生空位，从而使结构变得疏松，最终导致釉层脱落。

当水偏碱性时，则会对釉表面进一步产生腐蚀作用，碱性增强时反应会加快。

### （三）釉层在水中环境不同覆盖物下的表现

当陶瓷在水下环境中时，其表面覆盖的沉积物大致可分为两种：致密的沉积物和疏松的沉积物。致密的沉积物通常为颗粒细小的黏土软泥，疏松的沉积物为砾石、沙、贝壳、珊瑚碎片等。由于致密的沉积物层隔断了陶瓷与外界的物质交换，水也难于渗透，所以釉表面的风化作用难以进行，在这种情况下，陶瓷保存得较好。而在疏松的沉积物下，由于水流涌动带动沙粒等与釉面不断发生摩擦，对釉表面造成损害，同时，水对釉层的腐蚀在不同程度下进行，使釉层孔隙扩大，向纵深发展，甚至致使釉层剥落。

例如，南海一号沉船掩埋于珠江口海域一米以下的黏土质软泥中，该船所出瓷器光亮如新；而南海西沙和渤海湾中沉船的陶瓷，因其沉船遗址在粉砂以及砂与珊瑚碎屑交错的沉积物上，海水易渗透，加之海水涌动带动大量粉砂及珊瑚碎屑，与沉积层中的陶瓷不断发生摩擦，致使釉表面风化强度较高[98]。

### （四）釉层出水后的变化

陶瓷长期处于水下环境中，胎及釉的疏松部分均有水及可溶盐溶液渗入。当陶瓷出水后，原有的平衡被打破，析出的可溶盐结晶会对周围釉层产生压力，可能会导致釉表面发生开裂，甚至釉层剥落。

and Eastern China soil ranges from 5.5-6.5 pH; soil north of the Yangtze River is primarily neutral or basic, North and Northwestern China ranging from 7.5-8.5 pH and presenting with high amounts of $CaCO_3$; the highest recorded pH reaching as much as 10.5 [96].

## C. Underground Environment Weathering

### I. Weathering Caused By Liquids

Hydro-solutions reacting with the glaze surface are the primary causes of liquid weathering within the soil. The forms of weathering include: The $H^+$ and alkali metal ion exchange, producing voids and weakened structural integrity (see reaction ①); direct decomposition of the silica network by alkali bases (see reaction ②); or the deposit partially soluble salts and dissolving of soluble salts.

Water-soluble salts can enter into cracks, openings, and other glaze imperfections. As water evaporates, these salts crystallize. The repetitive dissolving and crystallization during the process of alternating moisture creates tension, and can crack the glaze, or even cause glaze erosion.

Both the soil and the glaze level contain many positive and negative particles ($Ca^{2+}$, $Mg^{2+}$, $Fe^{3+}$, $Mn^{2+}$, $Pb^{4+}$, $CO_3^{2-}$, $SO_4^{2-}$, $OH^-$, $SiO_3^{2-}$, $PO4^{3-}$), which can react with each other, creating varying colors of partially soluble salt deposits. It is possible for these partially soluble salts to become cemented to the surrounding soil, forming a hard dirt layer on the surface of the glaze.

Water solutions corrode the glaze surface, and develop a loosely porous geography across its face. At this time, if the surrounding soil particles enter into the glaze level, as the water evaporates, it leaves behind dirt deposits. The deposits' colors resemble the surrounding soil.

Due to the absorptive and porous capillary qualities of soil; water remains within the soil without flowing away. The mineral particles of soil are unevenly distributed, leading to different solutions and moisture in different areas. Therefore, a ceramic piece can present with varying severities and types of weathering across the glaze.

### II. Weathering Caused By Gasses

Air within soil is composed of essentially the same gasses as open air, with the main differences being:

a. Soil air exists in mutually isolated spaces within the soil; it is not a continuous system.

b. $O_2$ and $CO_2$ levels in soil are vastly different, with $CO_2$ being much higher than open air. Open air consist of 0.02%-0.03%, whereas the amount in soil is 0.15%-0.65% and as high as 5%. This is due primarily to the presence of organism respiration and the breakdown of organic materials. The amount of $O_2$ in soil is lower than that of open air, and moisture is much higher. Soil also contains trace amounts of reduced gasses such as: $CH_4$, $H_2S$, $H_2$, $NH_3$ and so forth [97].

Because of the heightened amounts of $CO_2$ and water in soil, associated weathering severity is drastically higher.

## D. Summary

As the underground environment is complex and heterogeneous, glaze corrosion is randomly distributed, with areas of severe weathering and areas retaining relatively original likeness.

Moist soil corrodes more severely than dry soil. Soils of alternating moisture corrode more severely than soils of consistent moisture. Acidic and basic soils corrode more than soils of pH neutrality. Differing from open-air weathering, the presence of partially soluble salts and soil particles in underground weathering can materialize yellow, brown, black and so forth discolorations.

If a piece is present within a relatively sealed and dry underground environment the weathering severity is lower than that of pieces that come into direct contact with soil.

# 3. Underwater Environment

## A. Underwater Environment Defined

Underwater Environment indicates a positioning situation in which a ceramic piece becomes emerged in a body of water.

## B. Underwater Environment Weathering

When a ceramic piece becomes submerged in water, the metallic ions (i.e. $Na^+$, $K^+$, $Pb^{2+}$) that are outside of the silica network enter into $H^+$ ion exchange, however the alkali formed become diluted in the water and cannot further corrode the glaze surface. If $R_2O$ is relatively high, ion exchange continues, therefore persistently increasing glaze surface porosity. In water, Pb ions continually migrate out and dissolve away. Over a long period of time, the glaze level becomes increasingly loose and eventually results in glaze erosion.

Water with alkalinity can move forward in glaze surface corrosion, with higher alkalinity increasing reaction speed.

## C. Glaze Level Under Different Material Covering

The material deposits that present on underwater ceramics can be divided into two main categories: Compact sediments and loose sediments. Compact sediments are clay and silt, while loose sediments are gravel, sand, shell fragments, and pieces of coral. Compact sediments can act as a partition inhibiting exchange between the glaze and external elements, as well as water permeation; because of this, it is difficult for weathering to proceed, and under this type of situation, the glaze is protected from corrosion. Whereas with loose sediments, the continuous contact of sand and other hard particles with the glaze face can cause damage. Additionally, water can continue corrosion to a certain degree, broadening and deepening glaze openings, even causing glaze erosion.

An example of this is Nanhai No. 1 shipwreck, which was buried under one meter of silt at the mouth of the Pearl River. The pieces excavated have a like-new sheen. Alternatively, in Xisha and Bohai Bay of the South China Sea, shipwreck pieces that were located among sand and coral debris, due to the contact with salt-water and constant friction caused by sand movement in underwater currents, presented with a higher weathering severity [98].

## D. Changes To Glaze Level After Water Excavation

Water and soluble salt solutions enter into the loosened glaze and body of ceramics that have spent a long period of time underwater. After excavation, the original equilibrium is disturbed, soluble salts crystallize, and stress is placed on the surrounding glaze; possibly causing glaze cracks, leading to glaze erosion.

# 第三节
# 釉层迹型风化特征

陶瓷釉层主要是由玻璃相、气相和晶相构成的，在经历岁月风化后，陶瓷迹型类别会发生各自不同的变化。一般来说，300年以上的陶瓷，都能在釉层上观察到以下所列的部分迹型变化现象。

## 一、玻璃相迹型风化特征

由于釉层玻璃相先天的凹凸不平以及釉层结构稳定性的差异，釉层经风化后，会在釉表面及釉层中的玻璃相上形成斑块和纹路两大迹型类别。

### （一）斑块

#### 1. 针眼斑

针眼斑是陶瓷在后天环境影响下，在釉表面以大小较均匀、深浅大致相同、疏密不定的聚集式独立微孔形式，呈不规则形貌随机分布的斑块。

#### 2. 坑点斑

坑点斑是陶瓷在后天环境影响下，在釉表面形成的，深浅不一、大小不一、形貌不一的不规则凹形斑块。

#### 3. 物盖斑

物盖斑是陶瓷在后天环境影响下，釉表面被外界物质长期覆盖而形成的斑块。

#### 4. 泡根斑

泡根斑是陶瓷在后天环境影响下，由破口泡发展形成的，内有沉积物的凹形斑块。

#### 5. 絮状斑

絮状斑是陶瓷在后天环境影响下，在釉表面及釉层中，由可溶盐、难溶盐和灰尘等物质沉积形成的，以白色为主，呈无序状，具有流淌漫延趋势的不规则斑块。

#### 6. 银屑斑

银屑斑是在后天环境影响下，在含铅较高的中、低温釉陶瓷的釉表面或釉层中形成的，由银白色沉积物构成的斑块。

### （二）纹路

#### 1. C形纹、O形纹

C形纹、O形纹是陶瓷在后天环境影响下，以字母C、O形貌分布在釉表面结构有缺陷处的线形纹路。

#### 2. 釉裂纹

釉裂纹是陶瓷在烧制过程中和在后天环境影响下形成的，以直线、曲线等不规则线条形貌分布在釉表面或釉层中的开裂纹路。

#### 3. 划痕

划痕是陶瓷釉表面与硬物碰划留下的破坏性纹路。

## 二、气相迹型风化特征

### （一）色变泡

色变泡是陶瓷在后天环境影响下，釉层中透明气泡的颜色发生了变化，但仍具透影性的气泡（透影性即半透明状态）。

### （二）壁变泡

壁变泡是陶瓷在后天环境影响下，釉层中的气泡内壁质地发生了风化现象，被风化处失去了透影性的气泡。

## 三、晶相迹型风化特征

### （一）针眼斑出现在流淌斑上

流淌斑（参见第六章第一节流淌斑）以含铁为主，容易风化。烧

# Section 3:
# Characteristics Of Weathered Glaze Level Traces

Ceramic glaze consists of glass, gas, and crystal phase components. After experiencing centuries of weathering, trace categories begin to experience different types of changes. Generally speaking, ceramics of three hundred years or more can present with observable properties of glaze level trace change phenomenon listed below:

## 1. Characteristics of Weathered Glass Phase Traces

After weathering, due to the innate uneven topography of the glaze level glass phase and differences in glaze level structural stability, the following two main categories: Mottles (Blotches) and Striae (Lines), can appear in the glass phase, across the glaze surface, or within the glaze level.

### A. Mottles

#### I. Needlepoint Mottles

Needlepoint Mottles are acquired mottles with a limited variation of size and depth, either amassed or independent, with randomly dispersed pits on the glaze surface, formed under the post-firing environment.

#### II. Pit Mottles

Pit Mottles (also called Pitting) are the irregularly shaped small depressions or pits on ceramic glaze, caused by external post-firing conditions, with non-uniform depth, size, and appearance.

#### III. Overlay Mottles

Overlay Mottles are acquired under the effect of the post-firing environment, and are the result of corrosion brought on by the long-term coverage of a foreign material over ceramic glaze.

#### IV. Bubble Roots

A Bubble Root (mottle) is depressed shape mottle with internal sediments, formed by the progression of a broken mouth bubble on the surface of ceramic glaze due to the influences of the post-firing environment.

#### V. Floccus Mottles

A Floccus Mottle is a primarily white colored appearance within the ceramic glaze level brought about by the post-firing environment; it has an irregular, free flowing, and disorderly shape that was formed by the depositing of soluble and partially soluble salts, dust particles, and other materials.

#### VI. Silver Chloride Mottle

Silver Chloride Mottles are the irregularly shaped silver-white colored deposits on low-temperature ceramic glazes with a relatively high amount of lead, caused by external post-firing conditions.

### B. Striae

#### I. C and O Shaped Striae

C and O shaped lines disbursed across the surface of ceramic glaze, in areas of structurally weakened glaze, acquired under the influence of the post-firing environment.

制成后，流淌斑无釉层保护的部分最先风化，形成针眼斑；有釉层保护的，在后天釉层受到侵蚀后也以针眼斑形式开始风化。

## （二）针眼斑出现在放射斑上

放射斑（参见第六章第一节放射斑）含一定量的铁氧化物，铁氧化物容易风化。烧制成后，无釉层保护的放射斑最先风化，形成针眼斑；有釉层保护的，在后天釉层受到侵蚀后也以针眼斑形式开始风化。

## （三）针眼斑出现在团状斑上

部分团状斑（参见第六章第一节团状斑）含一定量的铁氧化物，铁氧化物容易风化。大部分团状斑都处于釉层中。在后天釉层受到侵蚀后，也以针眼斑形式开始风化。

## （四）针眼斑出现在流淌纹上

流淌纹（参见第六章第二节流淌纹）含有一定量的铁，容易风化。没有被釉层保护的流淌纹以针眼斑形式开始风化，表现为针眼型阴纹。

### II. Glaze Cracks

Also called crazing, Glaze Cracks are any appearance of straight, curved, or irregular lines splitting the glaze surface of ceramics that was caused either during firing or under the influence of the post-firing environment.

### III. Scratches

Scratches are the imperfections on ceramic glaze that present as lines on the glaze surface that were left by the abrasive force of a solid external object.

## 2. Characteristics Of Weathered Gas Phase Traces

### A. Color Change Bubbles

Color Change Bubbles are the appearance of air bubbles within the glaze level that have undergone a coloration change under the influences of the post-firing environment, but have retained their translucency.

### B. Textured Bubbles

Textured Bubbles are the appearance of ceramic bubbles in which the interior has undergone weathering under the influence of the post-firing environment; gaining texture and losing their inherent translucency.

## 3. Characteristics Of Weathered Crystallized Traces

### A. Needlepoint Mottling On Flowing Mottles

Flowing Mottles (refer to Chapter 6 Section 1 Flowing Mottle) consist primarily of Iron, which under open air, easily corrodes. After firing, the areas without any glaze covering are the first to undergo weathering and form needlepoint mottles. Areas receiving protection from a thin glaze covering, when the glaze level encounters corrosion, can also present with needlepoint style weathering patterns.

### B. Needlepoint Mottling On Radiating Mottles

Radiating Mottles (refer to Chapter 6 Section 1 Radiating Mottle) contain certain amounts of Iron Oxide, which under open air, easily corrodes. After firing, the areas without any glaze covering are the first to undergo weathering and form needlepoint mottles. Areas receiving protection from a thin glaze covering, when the glaze level encounters corrosion, can also present with needlepoint style weathering patterns.

### C. Needlepoint Mottling On Blotch Mottles

A portion of Blotch Mottles (refer to Chapter 6 Section 1 Blotch Mottle) contain trace amounts of Iron Oxide, which exposed to air is easily weathered. The majority of blotch mottles are located within the glaze level. Areas receiving protection from a thin glaze covering, when the glaze level encounters corrosion, can also present with needlepoint style weathering patterns.

### D. Needlepoint Mottling On Flowing Striae

Flowing Striae (refer to Chapter 6 Section 2 Flowing Striae) consist primarily of Iron, which under open air, easily corrodes. After firing, the areas without glaze protection undergo weathering and form needlepoint mottling.

## 第四节
# 总结

陶瓷釉层的变化是釉层风化的结果，是釉层与多种物质发生物理变化、化学反应的综合结果。

在不同的存放环境中，无论是物理变化还是化学反应，其结果都会使陶瓷釉层迹型发生变化。陶瓷存放环境的变化，会使釉层的风化作用强度产生快慢的差异，从而导致釉层迹型风化强度有高有低。

这些迹型现象从一定程度上能够帮助我们推断陶瓷在后天存放环境的简单性或复杂性。

在长时间的风化过程中，风化产物在釉层的物理堆积方式和釉层的化学溶蚀结果上，一定与短时间人为的物理堆积和化学溶蚀结果有根本的区别。

# Section 4:
# Conclusion

Ceramic glaze level change is the total accumulated outcome of the physical changes and chemical reactions between the glaze level and various external materials.

Regardless of physical change or chemical reaction, the results are glaze level trace changes with varying characteristics. These weathering trace changes are heterogeneous. Changes in the post-firing environment bring forth differences in weathering intensity, thus the weathering severities experienced by the glaze level traces are similarly varied.

These trace change characteristics reflect the complexity of the ceramics' post-firing environment.

The aforementioned glaze level byproducts of physical amassing and chemical corrosion created over a long period of time, have fundamental differences to those resulting from immediate man-made physical amassing and chemical corrosion.

# 第六章
# 陶瓷典型迹型类别

**阅读提示：**

陶瓷釉层在烧制过程中及后天环境影响下，表现出各种迹型特征。陶瓷迹型会因釉的配方和工艺的不同，以及风化作用的差异而表现出不同的迹型特征。通过观察与研究，可将纷繁多样的陶瓷迹型归纳总结出不同类别。本章主要介绍陶瓷迹型中最典型的三大类别：斑块、纹路和气泡。为了更具体、更深刻地了解陶瓷迹型，本章选取元代青花瓷器的迹型类别作深入研究论述。

**主要论述问题：**

1. 陶瓷釉层中有哪些斑块种类？它们各自有什么特征？
2. 陶瓷釉层中有哪些纹路种类？它们各自有什么特征？
3. 陶瓷釉层中有哪些气泡种类？它们各自有什么特征？

## Chapter 6
# Typical Ceramic Trace Categories

**Text Note:**

Because of the firing process and post-firing environment, ceramic glazes present with several different varieties of traces. Additionally, due to differences in glaze recipes and craftsmanship, as well as the variances in weathering, several trace characteristic categories become evident. Through observation and research, we have been able to surmise and separate out the different categories of ceramic traces. The purpose of this chapter is to introduce the three main categories of ceramic traces: Mottles, Striations, and Air Bubbles. In order to offer a greater and more comprehensive look into these three categories, we have chosen Yuan Dynasty Underglaze Blue (Blue and White) Porcelain traces for deeper examination and clarification.

# 第一节
# 元青花斑块类别

## 一、元青花斑块概念

元青花斑块是元青花瓷器在烧制过程中和后天环境影响下形成的，以不规则形貌分布在釉表面及釉层中的迹型类别。

## 二、元青花斑块类型

如果将在烧制过程中形成的斑块称为先天性斑块，在后天环境影响下形成的斑块称为后天性斑块，那么元青花斑块就可以分成两大类型。

### （一）先天性元青花斑块

#### 1. 先天性元青花斑块概念

先天性元青花斑块是元青花瓷器在烧制过程中形成的，以不规则形貌分布在釉表面及釉层中的各类斑块。

#### 2. 先天性元青花斑块种类

先天性元青花斑块按照其形貌特征可分为三个种类：流淌斑、放射斑和团状斑。

（1）流淌斑

① 流淌斑概念

流淌斑是元青花瓷器在烧制过程中，在釉表面及釉层中形成的，以含铁为主，具有流淌感和不规则形貌的片状斑块。

流淌斑中的铁元素主要来源于青料。

② 流淌斑形成

元青花瓷器在高温烧制时，青料及釉料会逐渐熔化成熔融态，因浮力及气泡的搅动，胎、釉料以及青料中的铁氧化物在表面张力的作用下，在釉层中聚集，也可再形成分相。

铁氧化物易在气泡边缘聚集，部分会随气泡升至釉表面，顺着凹凸不平的釉表面向四周扩散，流动铺展，这就是一个润湿过程。

当温度下降时，这些铁氧化物便析出。由于它们的成分、密度、表面张力不同，在釉表面形成不同形貌、不同颜色的流淌斑。釉层中未升至釉表面的铁离子在高温下熔融在釉料中，釉层冷却凝固时在釉层中形成流淌斑。

③ 流淌斑分类

流淌斑可分为碘酒斑和沙滩斑两种。碘酒斑在釉层中受釉层保护；沙滩斑位于釉表面与空气接触。

A. 碘酒斑

碘酒斑是处于釉层中呈褐色的流淌斑，因与液体碘酒相似而得名，颜色有深有浅。由于碘酒斑已经和玻璃相融为一体，较难受到侵蚀，所以基本保持刚生产出的原态（见图6-1-1—图6-1-3）。

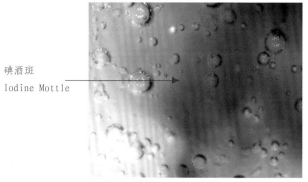

碘酒斑
Iodine Mottle

图6-1-1（N9）

# Section 1:
# The Yuan Dynasty Underglaze Blue Porcelain Mottle Category

## 1. YUB Mottles Defined

Yuan Underglaze Blue Mottles (spots and blotches) are a category of irregularly shaped appearances, created during the firing process and acquired under the post-firing environment, that are unevenly disbursed across the glaze surface and within the glaze level of Yuan Dynasty Underglaze Blue Porcelain (YUB).

## 2. YUB Mottle Classes

Mottles created during the firing process are called Innate Mottles. For those formed under the influences of the post-firing environment, the term Acquired Mottles is used. These are the two main classes of YUB Mottles.

### A. Innate YUB Mottles

I. Innate YUB Mottles Defined

Innate YUB Mottles are mottles of irregular shape and distribution on the glaze surface and within the glaze level of YUB porcelain that were formed during the firing process.

II. Innate YUB Mottle Types

According to the appearance characteristics of Innate Mottles, they can be split into three types: Flowing Mottles, Radiating Mottles, and Blotch Mottles.

a. Flowing Mottle

i. Flowing Mottle Defined

Primarily consisting of Iron, Flowing Mottles are the irregular blotch-shaped appearances across the glaze surface and within the glaze level which have a free flowing shape and were formed during the firing of YUB porcelain.

The Iron within Flowing Mottles comes primarily from the cobalt-blue dye, and has a considerably high concentration of Iron compared to other mottles.

ii. Formation of Flowing Mottle

During the firing of YUB, cobalt-blue dye and glaze gradually transitions into a molten state. Due to the buoyancy and movement of the air bubbles within the glaze level, the Iron Oxides of the body, glaze, and dye, under the effect of surface tension, accumulate together and can form into spilt phases.

It is easy for these Iron Oxides to accumulate around air bubbles and, a portion is then carried to and dispersed throughout the uneven surface of the glaze. This flowing dispersion is a fluid process.

When the temperature decreases, a portion of the Iron Oxides begins to crystallize. According to their composition, density, and surface tension differences; they present with alternate appearances and colors of flowing mottles across the surface of the glaze. Fe ions within the glaze, under high firing temperatures, melt into the glaze and form flowing mottles within the glaze level.

iii. Flowing Mottle Types

There are two types of flowing mottles: Iodine Mottles and Beach Mottles; with the iodine type occurring within the glaze level and receiving glaze protection, and the beach type occurring on the glaze surface in direct contact with the air.

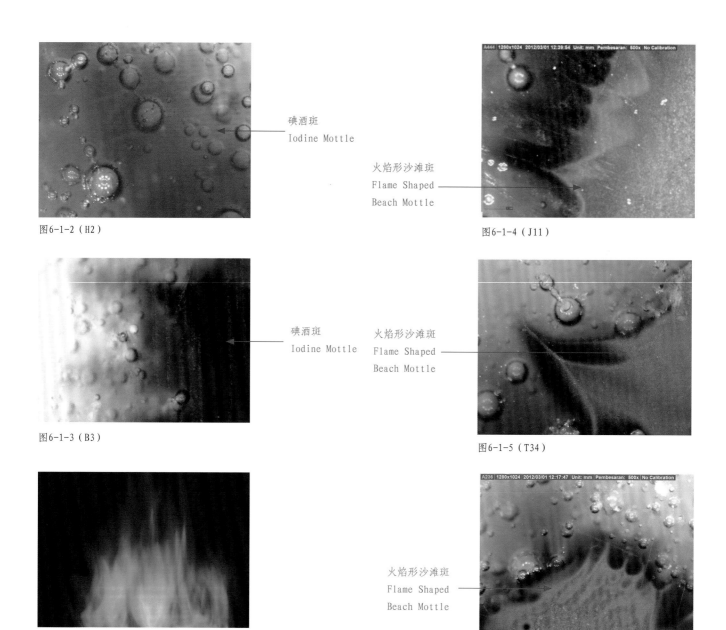

图6-1-2（H2）

图6-1-3（B3）

图6-1-7　真实火焰 Flame

图6-1-4（J11）

图6-1-5（T34）

图6-1-6（J4）

B. 沙滩斑

沙滩斑是呈褐色的流淌斑，因与沙滩相似而得名。沙滩斑基本都存在于釉表面。刚烧制出来的瓷器上的沙滩斑，由于它受釉层保护程度较低，与空气直接接触，所以最先风化。

风化后的沙滩斑，就是肉眼能看到的釉上黑斑，显微成像呈褐色。

沙滩斑主要有两种特征：一种表现为火焰形，一种表现为荷叶形。

火焰形沙滩斑斑块尾部呈火焰形，它是由于某部位的铁氧化物表面张力较小，流淌铺展形成的（见图6-1-4—图6-1-7[99]）。

荷叶形沙滩斑斑块尾部如同荷叶边沿，这类斑块的铁氧化物在流淌铺展时，表面张力比火焰形沙滩斑大（见图6-1-8—6-1-10）。

由于釉表面的凹凸不平，有时沙滩斑上会出现不规则圆孔，分布如同筛网。有的沙滩斑在流淌润湿时无法完全覆盖凹凸不平的釉表面，留下凸起的圆孔（见图6-1-8—6-1-10）。

一般来说，荷叶形沙滩斑上出现圆孔的几率要比火焰形沙滩斑要高。

见图6-1-11、图6-1-12（图6-1-11、图6-1-12为同一器物同一位置图片，图6-1-11为二维图片，图6-1-12为三维图片）。

箭头A所指的圆孔所处位置高

(a) Iodine Mottle

Iodine Mottles are iodine colored flowing mottles that exist within the glaze level and present on the fringe of flowing mottles with a color spectrum from deep to light. Due to their position within the glaze level, they have fused within the glass phase and are, as such, resistant to corrosion; therefore they preserve the fundamental conditions attained during the firing process (see pictures 6-1-1—6-1-3).

(b) Beach Mottle

Beach Mottles are flowing mottles that have a sandy beach appearance and are located on the surface of the glaze. Newly formed beach mottles rise up above the surrounding glaze surface, and they present in the center of flowing mottles. Due to its direct contact with the surrounding air, and without glaze protection it is highly susceptible to weathering.

This exposed portion of the mottle is what is visible to the naked eye, called 'tiexiu ban' or 'heap and piling', and presents with a brownish color under magnification.

Beach mottles have two main types of characteristics:
The first type, appearing as flame shaped;
The second type, appearing as lotus leaf shaped.

Flame shaped beach mottle edges resemble a flickering flame; this is the result of low surface tension that allows the flow across the glaze surface, creating flame shaped edges on the beach mottle (see pictures 6-1-4—6-1-7 [99]).

Lotus leaf shaped beach mottle edges resemble the edge of a lotus leaf. When the mottles flow across the surface of the glaze, their surface tensions are normally higher than flame shaped beach mottles (see pictures 6-1-8—6-1-10).

Due to the uneven surface of the glaze. Beach mottles that have furrows and troughs across the surface create a pattern similar to that of a sieve. Some of the beach mottles are unable to completely cover the uneven glaze surface, leaving behind furrows of glaze (see pictures 6-1-8—6-1-10).

Generally speaking, lotus leaf shaped beach mottles are more likely to have furrows and troughs than flame shaped beach mottles.

Look at pictures 6-1-11 and 6-1-12 (where 6-1-11 is the two-dimensional and 6-1-12 is the three-dimensional view of the same spot on the same piece).

Arrow A is pointing at a raised furrow within the beach mottle. The Iron Oxide, during the fluid process, was unable to submerge the furrow. Because the furrow is glass phase, the degree of weathering is less evident and able to essentially retain its original form.

Look at pictures 6-1-13 and 6-1-14 (where 6-1-13 is the two-dimensional and 6-1-14 is the three-dimensional view of the same spot on the same piece).

Arrow B is pointing to the beach mottle and Arrow C is pointing at the glaze surface glass phase. Point B is comparatively lower than that of point C. This is because the relatively high amount of Iron Oxide at point B, after weathering, has difficulty adhering to the glaze surface and is easily corroded away. After a long period of time and weathering decay, a gradual depression occurs at point B of the mottle.

b. Radiating Mottle

i. Radiating Mottle Defined

Formed during the firing of YUB porcelain, Radiating Mottles are the irregular blotch shaped appearances across the glaze surface and within the glaze level, which have a crystallized, radiating form.

ii. Formation of Radiating Mottle

During the firing of YUB, cobalt-blue dye and glaze gradually transitions into a molten state. Due to the buoyancy and movement of the air bubbles within the glaze level, the Iron Oxides of the body, glaze, and dye, under the influence of surface tension, accumulate together and can form into spilt phases. It is easy for these Iron Oxides to accumulate around air bubbles, and a portion is then carried to the surface of the glaze.

When the temperature decreases, a portion of the Iron Oxides begins to crystallize. According to their composition, density, and surface tension differences, they present with alternate appearances and colors of radiating mottles across the surface of the glaze or within the glaze level.

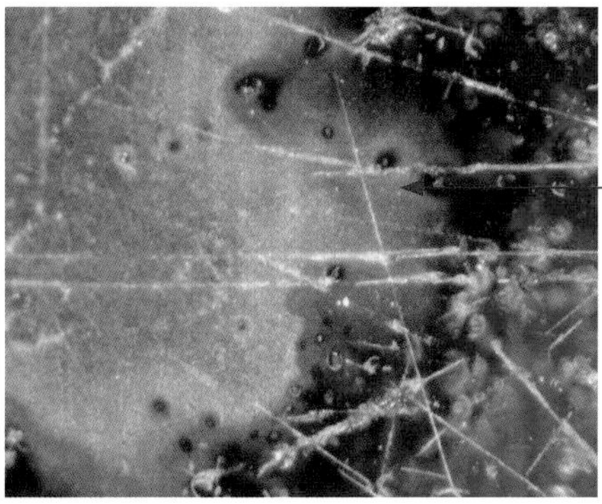

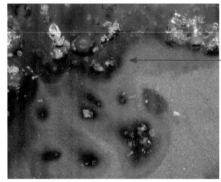

荷叶形沙滩斑
Lotus Leaf shaped
Beach Mottle

图6-1-8（T14）

A

图6-1-11（B2）

荷叶形沙滩斑
Lotus Leaf shaped
Beach Mottle

图6-1-9（T34）

A

图6-1-12（B2）

A132 1280x1024 2012/03/01 12:07:10 Unit: mm Pembesaran: 500x No Calibration
荷叶形沙滩斑
Lotus Leaf shaped
Beach Mottle

图6-1-10（J3）

B
C

图6-1-13（DY7）

B
C

图6-1-14（DY7）

出沙滩斑，铁氧化物在流淌润湿的过程中，没有将凸起的圆孔淹没。圆孔处属玻璃相，风化作用不明显，因此基本保持原有形貌。

见图6-1-13、图6-1-14（图6-1-13、图6-1-14为同一器物同一位置图片，图6-1-13为二维图片，图6-1-14为三维图片）。

箭头B所指的沙滩斑与箭头C所指的釉表面玻璃相相比凹陷下去了，这是因为沙滩斑含铁氧化物，铁氧化物风化后，不易附着在釉表面，较易脱落。在长时间的风化与脱落后，沙滩斑逐渐变成凹坑。

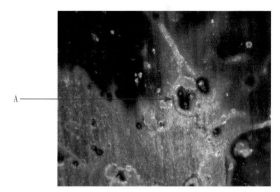

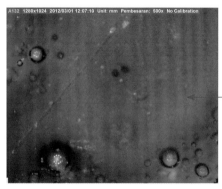

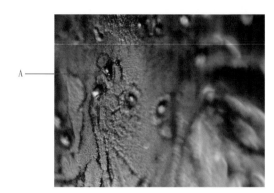

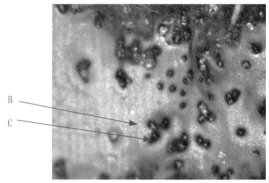

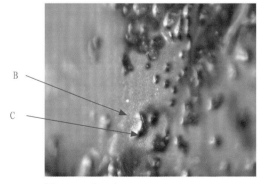

iii. Radiating Mottle Types

Radiating Mottles can be divided into five color series: white series, red series, yellow series, brown series and black series.

Almost all radiating mottles have certain amounts of Iron Oxide, white series being the exception with little or none. Due to the differences in Iron valance, composition, and density, as well as the degree to which they receive glaze protection, these mottles present with various colors and appearances.

White series mottles appear as either a chrysanthemum-like shape or in pine needle bunches, and contain little or no Iron Oxides (see pictures 6-1-15— 6-1-18).

Red series mottles primarily consist of rust and tangerine color and usually present in a maple leaf shape (see pictures 6-1-19 and 6-1-20).

Yellow series mottles present with a color range from light to dark yellow and have irregular shape, these mottles primarily occur within the glaze level with a high degree of glaze protection (see pictures 6-1-21—6-1-27) Some Yellow series radiating mottles present with black split phase crystallization at their peak (see pictures 6-1-23 and 6-1-24).

Brown series mottles have a coffee color and are irregular in shape, primarily occurring at the surface of the glaze with a low degree of glaze protection (see pictures 6-1-26— 6-1-29).

Black series mottles are black in color and are of irregular shape, primarily occurring at the surface of the glaze with a low degree of glaze protection (see pictures 6-1-28 and 6-1-29).

Most radiating mottles form in Iron Oxide rich areas of dense cobalt-blue dye, such as areas of underglaze blue pattern (see pictures 6-1-24 and 6-1-25); or in conjunction with flowing mottles (see pictures 6-1-19 — 6-1-23).

Similar to flowing mottles, portions of radiating mottles occur within the glaze level while others occur on the surface, uncovered by glaze. After the firing and solidifying of the glaze, the yellow series mottles that are not covered by glaze are exposed to the air, forming brown or black series mottles, while those that are protected by the glaze are less susceptible to weathering and able to retain their original color, remaining within the yellow series (see pictures 6-1-26— 6-1-29).

As radiating mottles belong to the crystal phase, they are different to that of the glass phase of glaze. This phase difference as well as discrepancy in surface tension disallows the fusion of these two bodies on the glaze surface, creating a difference in topography between the two phases. Some mottles are higher than the surrounding glaze level, and some are lower. Below are two examples.

Mottle higher than the surrounding glaze level: See the radiating mottle indicated by arrow D in pictures 6-1-30 and 6-1-31 (where picture 6-1-30 is the two-dimensional and 6-1-31 is the three-dimensional view of the same spot on the same piece).

Mottle lower than the surrounding glaze level: See the radiating mottle indicated by arrow E in pictures 6-1-32 and 6-1-33 (where picture 6-1-32 is the two-dimensional and 6-1-33 is the three-dimensional view of the same spot on the same piece).

c. Blotch Mottle

i. Blotch Mottle Defined

Formed during the firing of YUB porcelain, Blotch Mottles are those within the glaze level that have assembled into one group but do not possess a radiating shape. They are a crystallized formation.

ii. Formation of Blotch Mottle

Blotch mottles are similar to the pine needle radiating mottles discussed earlier in that they are both formed by the crystallization of porcelainite (or anorthite). Pine needle and blotch mottles are rarely seen within YUB, due in part to the changes in glaze recipe and techniques.

Research by Chen Yaocheng et al goes to show [100]: Only if the amount of Aluminum Oxide in the dye exceeds 18%-20%, the fusion of dye and glaze can occur to create porcelainite or anorthite crystallizations. During high temperature firing, while using Aluminum-rich blue dye to color underglaze blue porcelain, only a portion of the blue dye melts to create rich Aluminum blue glass phase. During the cooling, at the dye-glaze fusion points, porcelainite and anorthite crystals can

（2）放射斑

① 放射斑概念

放射斑是元青花瓷器在烧制过程中，在釉表面及釉层中形成的，具有放射感的析晶斑块。

② 放射斑形成

元青花瓷器在高温烧制时，青料及釉料会逐渐熔化成熔融态，因浮力及气泡的搅动，胎、釉料以及青料中的铁氧化物在表面张力的作用下，在釉层中聚集，也可再形成分相。

铁氧化物易在气泡边缘聚集，部分会随气泡升至釉表面。当温度下降时，这些铁氧化物便析出。由于它们的成分、密度、表面张力不同，在釉表面、釉层中形成不同形貌、不同颜色的放射斑。

③ 放射斑分类

放射斑可分成五大色系：白色系、红色系、黄色系、褐色系及黑色系。

除白色系放射斑不含或含极少铁氧化物之外，大部分放射斑都一定程度地含有铁氧化物。由于铁价位、结构及浓度的不同，加之受釉层保护程度不同，这些放射斑呈现颜色及形貌差异。

白色系中有菊花形放射斑及松针型放射斑（见图6-1-15—图6-1-18）。

红色系中以铁红和橘红色为主，常见有枫叶形（见图6-1-19、图6-1-20）。

黄色系中包括淡黄、深黄等颜色，属于不规则形貌一类，通常出现在釉层中，受釉层保护程度较高（见图6-1-21—图6-1-27）。

图6-1-15（Z6）

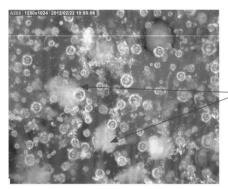

图6-1-16（Z6）

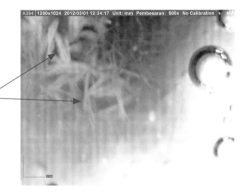

图6-1-17（J8）

图6-1-18（J15）

有的黄色系放射斑上还有黑色系分相结晶析出（见图6-1-23、图6-1-24）。

褐色系中主要是咖啡色，属于不规则形貌一类，通常出现在釉表面，受到的釉层保护程度较低（见

form. Because the majority of YUB samples do not have an Aluminum content exceeding 18%, these types of mottles are rarely seen; however, they become much more common on early Ming Dynasty pieces.

Additionally, since porcelainite crystallization is directly impacted by firing temperature, the presence of porcelainite indicates a firing temperature of over 1000ºC [101]. As such, there is a strong possibility that pieces presenting with blotch mottling were fired during or after the late Yuan Dynasty period.

iii. Blotch Mottle Properties

Blotch mottles appear in the shape of smooth stones, with white and yellow being the primary colors, and with varying levels of opacity. Some blotch mottles present with a single color (see picture 6-1-34) while others present with a color spectrum (see picture 6-1-35).

d. Mixed Mottle

Mixed Mottles are the different combinations of radiating, flowing, and/or blotch mottles on the same piece of porcelain.

i. Mixed Mottle Grouping

It is common to see the combination of flowing and radiating mottles on a single piece of YUB porcelain.

Normally, the most commonly seen flowing and radiation mottle combinations are: Iodine, beach and brown or black series radiating mottle combination (see pictures 6-1-36 and 6-1-37). This is the most common YUB mottle group; also seen is the iodine and radiating mottle combination (see picture 6-1-38); with the iodine, beach and red series radiating mottle combination (see picture 6-1-39) being the least common.

Flowing and blotch mottles are also rarely seen, because the limited amount of blotch mottles on YUB porcelain. If seen, it is most commonly the combination of iodine and blotch mottles (see picture 6-1-40).

In the same manner, radiating and blotch mottles can be considered very rare on YUB porcelain. If seen, they primarily consist of the combination of pine needle radiating and blotch mottles (see pictures 6-1-41 and 6-1-42).

These mixed mottle combinations can occur in conjunction, overlapping, or in complement to one another. However, no matter the combination or relationship with surrounding mottles, there is a clear delineation amongst each mottle.

ii. Mixed Mottle Properties

Beach and radiating mottle combinations are separated into two categories: 'Mottle within Mottle' and 'Mottle on Mottle'.

Mottle within Mottle: The surrounding beach mottle encompasses the radiating mottle with a glass phase border separating the two mottles (see pictures 6-1-43 and 6-1-44).

Mottle on Mottle: Radiating and beach mottles overlap with no visible glass phase boundary (see pictures 6-1-45 and 6-1-46).

It is also possible for 'mottle on mottle' and 'mottle within mottle' to occur simultaneously (see pictures 6-1-47 and 6-1-48).

e. Summary

The YUB porcelain body and glaze both have trace amounts of elemental Iron, but the amount of Iron in the cobalt-blue dye is higher. There are different types and appearances of crystallization mottles; some spilling over onto the glaze surface and some existing only within the glaze level. Most of them occur in areas of dense blue dye, with the majority containing a certain amount of Iron Oxides.

There are multiple appearances and colors of Iron-rich crystallization mottles, which are formed dependent on the viscosity and thickness of the glaze as well as the surface tension, density, and elemental Iron positioning of the split phase components.

Note: All microscopic pictures of innate characteristics on YUB porcelain samples have undergone weathering; all of the observable qualities of innate mottles have experienced approximately eight hundred years of history. Even the most protected pieces of porcelain have undergone certain amounts of weathering. As we do not have the capability of entering back into the unique YUB porcelain environment, we can only use the pictures gained through microscopic observation

红色系枫叶形
放射斑
Red Series Maple
Leaf Shaped
Radiating Mottles

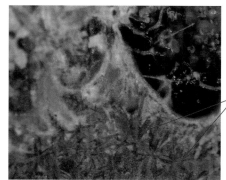

红色系枫叶形
放射斑
Red Series Maple
Leaf Shaped
Radiating Mottles

图6-1-19（B3）　　　　　　　　　　图6-1-20（T30）

黄色系
放射斑
Yellow Series
Radiating Mottle

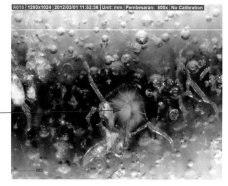 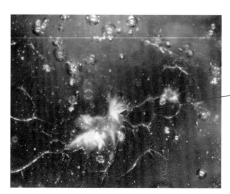

黄色系
放射斑
Yellow Series
Radiating Mottle

图6-1-21（J1）　　　　　　　　　　图6-1-22（N2）

黄色系
放射斑
Yellow Series
Radiating Mottle

黑色系分相结晶
Black Series Split
Phase Chystallization

 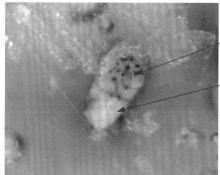

黑色系分相结晶
Black Series Split
Phase Chystallization

黄色系放射斑
Yellow Series
Radiating Mottle

图6-1-23（H1）　　　　　　　　　　图6-1-24（N5*）

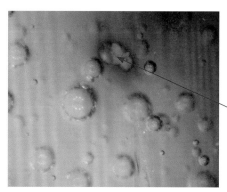

黄色系放射斑
Yellow Series Radiating Mottle

图6-1-25（H1）

*图片经过裁剪放大处理。
*Picture cropped and enlarged to show detail.

in addition to current scientific knowledge to postulate the appearance and characteristics of the original inherent qualities.

## B. Acquired YUB Mottle

### I. Acquired YUB Mottle Defined

Acquired YUB Mottles are mottles of irregular shape and distribution on the glaze surface and within the glaze level of YUB porcelain that were formed under the influences of the post-firing environment.

### II. Acquired YUB Mottle Types

#### a. Needlepoint Mottle

##### i. Needlepoint Mottle Defined

Needlepoint Mottles are YUB acquired mottles with a limited variation of size and depth, either amassed or independent, with randomly dispersed pits on the glaze surface, formed under the post-firing environment.

Normally, needlepoint mottles are first formed in areas on crystallization mottles without glaze coverage, in areas of glaze defect, or areas covered by soil or sand.

##### ii. Needlepoint Mottle Types

Needlepoint mottles occur in random distribution, and can occur anywhere on the glaze surface; they can be divided into three types of appearances:

(a) Foggy Needlepoint Mottle

Foggy needlepoint mottles are spots of a finely distributed misting of pits giving the appearance of fog. Usually, their boundaries have a clear progression from dense to sparse, and present on areas of exposed Iron Oxide (see pictures 6-1-49 — 6-1-52).

(b) Frosted Needlepoint Mottle

Frosted needlepoint mottles are simply those with larger needle pricks than that of foggy needlepoint mottles, with a distribution density and weathering severity higher than that of the foggy variety, giving the appearance of sand-blasted or frosted glass. They can be further divided into two types:

The first type is the progression from foggy needlepoint mottles, with a high density and amount of weathering, but also presenting on areas of exposed Iron Oxide (see pictures 6-1-53 — 6-1-55).

The second type occurs in areas where Overlay Mottles have been cleaned away from the glaze surface, leaving frosted needlepoint mottles. For clarity sake, we refer to this type as overlay frosted needlepoint mottles. This type has random distribution, irregular shape, and fine texture; normally not presenting with a density progression, but still retaining clearly delineated boundaries similar to that of coastlines (see pictures 6-1-56 and 6-1-57).

(c) Grainy Needlepoint Mottle

Normally, grainy needlepoint mottles are those with even bigger and deeper pits than that of foggy and frosted needlepoint mottles, with a weathering severity even higher than that of the foggy and frosted variety, and large clearly delineated grains. The distribution density is usually lower than that of frosted needlepoint mottles, but can similarly be further divided into two types:

The first type also present on exposed Iron Oxidized areas with little or no glaze (see picture 6-1-58) or in areas of glaze defect (see picture 6-1-59).

The second type occurs in areas where Overlay Mottles have been cleaned away from the glaze surface, leaving grainy needlepoint mottles. For clarity sake, we refer to this type as overlay grainy needlepoint mottles (see picture 6-1-60).

##### iii. Distribution of Needlepoint Mottling

(a) Needlepoint Mottling on Flowing Mottles

(i) Formation of Needlepoint Mottles on Flowing Mottles

Flowing mottles consist primarily of Iron, which under open air, easily corrodes. After firing, the areas without any glaze covering are the first to undergo weathering and form needlepoint mottles. Areas receiving protection from a thin glaze covering, when the glaze level encounters corrosion, can

图6-1-26—图6-1-29）。

黑色系中主要是黑色，属于不规则形貌一类，通常出现在釉表面，受到的釉层保护程度较低（见图6-1-28、图6-1-29）。

凡含铁氧化物的放射斑几乎都出现在青料浓处，这包括青花纹饰浓处（见图6-1-24、图6-1-25）或流淌斑处（见图6-1-19—图6-1-23）。

和流淌斑一样，有的放射斑受釉层保护的程度不一。在烧制过程中，瓷器釉层冷却凝固后，处于釉表面受釉层保护程度较低的黄色系放射斑斑块顶部，因与空气接触而形成褐色系或黑色系放射斑，而处于釉层中的那部分斑块受釉层保护程度较高，仍保持黄色（见图6-1-26—图6-1-29）。

放射斑属于晶相，与釉层玻璃相属于两种不同的相，在交接面汇

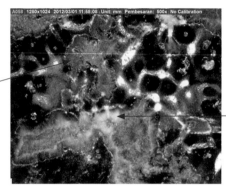

图6-1-26（J1）

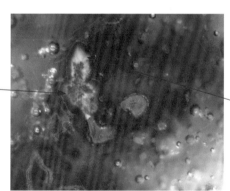

图6-1-27（H4）

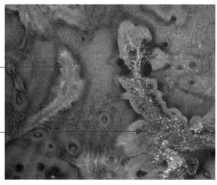

图6-1-28（H5）

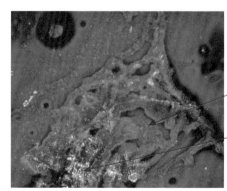

图6-1-29（N5）

also present with needlepoint style weathering patterns.

(ii) Properties of Needlepoint Mottling on Flowing Mottles

Needlepoint mottles appear white due to their refractive properties and are dispersed over flowing mottles that are at the glaze surface (beach mottles), with some areas dense and some areas sparse. In general, the needlepoint mottling does not extend past the boundaries of the flowing mottle.

Some foggy needlepoint mottles have the distribution appearance of spray-paint over the top of the glaze, presenting with evenly thinning edges. There is a lack of clearly delineated boundary with the glass phase, but rather a well-distributed progression from dense to sparse (see picture 6-1-61).

Frosted needlepoint mottles are often most dense towards the central portions of the flowing mottle due to less glaze level protection and higher weathering severity at that point; the closer to the beach mottle boundary, the more protection offered by the glaze, and the sparser the distribution of needlepoint mottling; eventually fading into the foggy variety (see picture 6-1-62).

Frosted needlepoint mottling usually does not have a spectrum from thick to thin, but is more delineated, helping to give a layered effect to beach mottles.

(b) Needlepoint Mottle on Radiating Mottles

(i) Formation of Needlepoint Mottles on Radiating Mottles

Radiating mottles contain certain amounts of Iron Oxide, which under open air, easily corrodes. After firing, the areas without any glaze covering are the first to undergo weathering and form needlepoint mottles. Areas receiving protection from a thin glaze covering, when the glaze level encounters corrosion, can also present with needlepoint style weathering patterns.

(ii) Properties of Needlepoint Mottles on Radiating Mottles

Due to the increased complexity of surface texture in radiating mottles compared to that of flowing mottles, the appearance of needlepoint mottling on radiating mottles is accordingly more complex.

Needlepoint mottles appear white due to their refractive properties and are dispersed over radiating mottles that are at the glaze surface, with some areas dense and some areas sparse.

A portion of foggy needlepoint mottling extends across the glass phase up to the boundary of the radiating mottles, with a clearly delineated border, and very limited amount of weathering occurring on the radiating mottle. This in turn creates an "outside full, inside empty" phenomenon (see picture 6-1-63).

Conversely, there is a portion of foggy needlepoint mottling that occurs within the boundary of the radiating mottle, creating a clear delineation with the surrounding glass phase. This, in turn, creates an "inside full, outside empty" phenomenon. In general, foggy needlepoint mottling occurs in areas that have a certain amount of glaze covering; areas of lower weathering severity. This creates a very fine mist covering over the radiating mottle, and allows us the ability to observe the original likeness of the radiating mottle through the fog (see picture 6-1-64).

A portion of frosted needlepoint mottling occurs in conjunction with the shape of radiating mottles, often presenting at the most central portion of the mottle, as white speckles, without extending past the surroundings of the radiating mottle, and a lacking spectrum of density (see pictures 6-1-65 and 6-1-66).

Some radiating mottles have a surrounding border, which is the boundary between two dissimilar phases. There is inherent structural weakness in these inter-phase boundaries and troughs form, which allows them to weather more easily.

This encompassing border assumes a white or yellow color under magnification, and is thin and fine with only a small portion accumulating substantial width. Overall, they are fine and restrained, yet distinct. They more or less follow the shape of the radiating mottle, while retaining a slight distance from the edge of the mottle (see picture 6-1-67).

(c) Needlepoint Mottles on Blotch Mottles

(i) Formation of Needlepoint Mottle on Blotch Mottles

A portion of blotch mottles contain trace amounts of Iron Oxide, which exposed to air is easily weathered. The

合时这两相是不相融的，由于它们的表面张力不同，两相之间会有凹凸不平的情况，有时是斑块高出釉表面，有时是斑块低于釉表面。

斑块高出釉表面，见图6-1-30、图6-1-31中箭头D所指放射斑（图6-1-30、图6-1-31为同一器物同一位置图片，图6-1-30为二维图片，图6-1-31为三维图片）。

斑块低于釉表面，见图6-1-32、图6-1-33中箭头E所指放射斑（图6-1-32、图6-1-33为同一器物同一位置图片，图6-1-32为二维图片，图6-1-33为三维图片）。

（3）团状斑

① 团状斑概念

团状斑是元青花瓷器在烧制过程中，在釉层中形成的聚集成一团、没有放射感的析晶斑块。

② 团状斑形成

团状斑与放射斑中的松针形放射斑类似，都属于莫来石（或钙长石）的析晶。松针形放射斑与团状斑在元青花上都较少见，这与釉料配方变化及工艺改变有关。

陈尧成等人的研究表明，如果青料中的氧化铝含量超过18%—20%，釉与青料结合处的成分才有可能析出钙长石或莫来石结晶。当用含高铝青料着色时，烧成的青花瓷器只有一部分青料被釉熔融形成富铝蓝色玻璃相，冷却时釉与青料结合处就会有钙长石或莫来石相析出。而大部分元青花标本的青料中氧化铝含量都不及18%，所以元青花瓷器上少有这类析晶斑，而在明早期青花瓷器上较常见[100]。

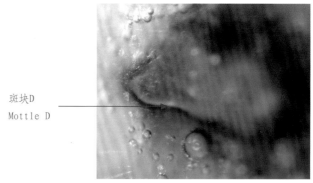

图6-1-30（B3）

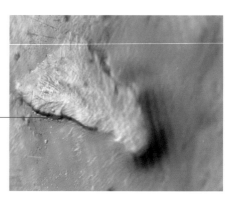

图6-1-31（B3）

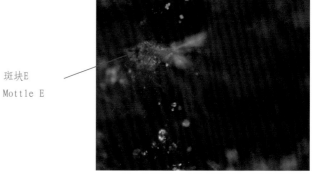

图6-1-32（B3）

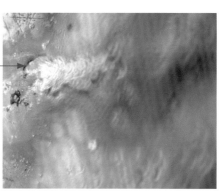

图6-1-33（B3）

majority of blotch mottles are located within the glaze level. Areas receiving protection from a thin glaze covering, when the glaze level encounters corrosion, can also present with needlepoint style weathering patterns.

(ii) Properties of Needlepoint Mottling on Blotch Mottles

Because large portions of blotch mottles have a low concentration of Iron or are protected within the glaze level, the degree of weathering is relatively weak, presenting with only a sparse amount of foggy needlepoint mottling. Resembling a very thin film covering over the blotch mottle, the presence of needlepoint mottling does not inhibit the ability to observe the original characteristics of the blotch mottle (see picture 6-1-68).

Similar to the aforementioned radiating mottle situation, blotch mottles can present with a very fine needlepoint mottle border; it is restrained, yet distinct. The borders roughly follow the shape of the mottle, while retaining a slight distance from its edge (see picture 6-1-68).

(d). Needlepoint Mottles on Glass Phase

(i) Formation of Needlepoint Mottles on Glass Phase

Needlepoint Mottling on glass phase glaze occurs in two different types:

Areas on the glaze surface with exposed structural flaws; for example areas of split phasing are easily weathered and needlepoint mottling can occur;

Areas of the glaze surface that have undergone a long period of foreign material covering (i.e. dirt, soot, soil, etc.), when cleared away, often leave needlepoint mottling at the point of foreign material coverage. This type of needlepoint mottling is corrosion caused by the foreign material.

(ii) Properties of Needlepoint Mottling on Glass Phase

Normally, needlepoint mottles present as white due to their refractive nature.

Grainy Needlepoint Mottles: On the glaze surface, areas of weakness form grainy needlepoint mottles (see picture 6-1-69); or areas in which Iron Oxide is exposed (see picture 6-1-70).

Overlay Frosted Needlepoint Mottles: There are clearly delineated boundaries between the needlepoint mottling on glass phase brought about by the overlay of foreign materials and the glass phase that has not been weathered (see picture 6-1-71).

Overlay Grainy needlepoint mottles: They are dispersed independently with clear distinction between each weathered point and the unweathered portions of glass phase that surrounds them (see picture 6-1-72).

b. Pit Mottle

i. Pit Mottle Defined

Pit Mottles (also called Pitting) are the irregularly shaped small depressions or pits on YUB porcelain glaze, caused by the external post-firing environment, with non-uniform depth, size, and appearance.

ii. Formation of Pit Mottles

There are four main causes of Pit Mottles:

(a) The progression of 'C' and 'O' shaped striae (refer to Section 2 of this chapter, C and O shaped striae): As C and O shaped striae continue to weather, they form pit mottles; as pit mottles continue to weather, they can chain together into pit groups.

(b) The progression of needlepoint mottling: The further weathering of needlepoint mottles changes into pitting.

(c) Damage caused by external collision: This type of pitting is usually larger in surface area, often presenting with further needlepoint mottling in the center.

(d) Overlay Material Corrosion: Once foreign materials are washed away, pitting can present.

iii. Properties of Pit Mottles

As some pitting occurs via the progression of C and O shaped striae, it is common to see C and O shaped striae in conjunction with this form of pitting. This is considered as the most typical case of pitting (see pictures 6-1-73 — 6-1-75), and is a reflection of the passage of time.

Initially, pit mottles are dispersed independently across the glaze surface, but as weathering continues, they can become

此外，莫来石结晶与烧造温度也有关，有莫来石结晶说明烧结温度至少在1000℃以上[101]，以此可以推测，这类元青花瓷器很有可能产于元代末期。

③ 团状斑特征

团状斑形貌如石块，颜色主要以白色、黄色为主，部分斑块具备透影性，部分不具备透影性。部分斑块颜色单一（见图6-1-34），部分有色阶变化（见图6-1-35）。

（4）混合斑

混合斑是放射斑、团状斑及流淌斑，经不同组合，同时存在于同一瓷器上的斑块现象。

① 混合斑组合

在同一元青花瓷器上，常见的组合有流淌斑与放射斑。

一般情况下，流淌斑与放射斑的组合更常见：主要是碘酒斑、沙滩斑与褐色系或黑色系的放射斑为主的组合（见图6-1-36、图6-1-37），这是最常见的斑块组合；也有以碘酒斑与放射斑为主的组合（见图6-1-38）；碘酒斑、沙滩斑与红色系的放射斑的组合较少见（见图6-1-39）。

流淌斑与团状斑的组合较少，因为团状斑本身在元青花上就比较罕见，而这种组合中的流淌斑常以碘酒斑形式出现（见图6-1-40）。

放射斑与团状斑的组合也较少见，主要有松针形放射斑与团状斑组合（见图6-1-41、图6-1-42）。

这些组合的存在形貌有时并

图6-1-34（J15）

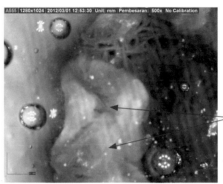

图6-1-35（J15）

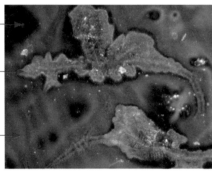

图6-1-36（B3）

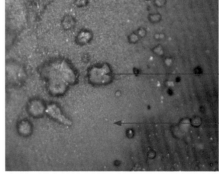

图6-1-37（N12）

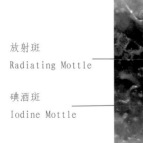

图6-1-38（J1）

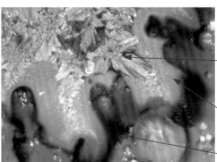

图6-1-39（B3）

linked with surrounding pits and form spiraled pit groups, often assuming an irregular shape (see pictures 6-1-76 — 6-1-78).

According to the depressed shape of pit mottles, it is easy for sediments to become deposited along the pit floor, and present with some color spectrum (see picture 6-1-79).

Pitting that occurs via the progression of needlepoint mottling is randomly dispersed and will assume white coloration as well as varying sizes. They often present with a fine, winding border; with needlepoint mottling still visible within the greater encompassing pit mottle (see picture 6-1-80). Others present with dense needlepoint mottling within the pit. They are dense to the point that the needlepoints have lost independent shape, thus opacity is achieved (see picture 6-1-81).

The difference between needlepoint mottles and pit mottles that developed from needlepoint mottling is:

Pit mottles are independent entities with non-uniform size, depth, appearances and dispersion; the glaze level is visible between pit mottles; and the boundaries between pits and the surrounding glaze are very distinct (see picture 6-1-82).

Additionally, the case with the majority of needlepoint mottles is that the glaze level base is not visible between the needlepoints, similar to the spread pattern of a spray paint blast. Even and sparsely dispersed needlepoint mottling, while the glaze level may be visible, the size and depth of the mottling is uniform (see picture 6-1-83).

Pit mottles caused by external abrasion or collisions are often larger in size, with some extending all the way to the clay body. The sedimentary deposits and needlepoint mottles within the pit are visible and can assume a wide array of colors (see picture 6-1-84).

After a long period of time in an environment of varying moisture it is possible for pitting to also include floccus mottling (refer to Chapter 6 Section 1 Floccus Mottle) (see picture 6-1-85).

It is also possible to have different types of pitting occur in conjunction with one another (see picture 6-1-86).

c. Overlay Mottle

i. Overlay Mottle Defined

Overlay Mottles are acquired under the effect of the post-firing environment, and are the result brought on by the long-term coverage of a foreign material over YUB porcelain glaze.

ii. Formation of Overlay Mottles

Majority of YUB overlay mottles come about because the piece has been stored for a period of time underground or within a wet environment. They consist of all the dirt and other materials that have covered the surface of the porcelain for an extended period of time, as well as any manmade or foreign materials added to repair the piece.

iii. Properties of Overlay Mottles

The primary colors of overlay mottles are yellow and brown or greyish-white with a clear delineation from the surrounding glaze surface. Unlike needlepoint mottling with irregular edges, it can have straight or irregular edges as well as having distinct layers.

It can present as peeling style (see picture 6-1-87), with mud cracks (see picture 6-1-88) much like irrigated farmland, and can have scrapes across the mottle (see picture 6-1-89).

Some overlay mottles, through long term corrosive contact to the glaze surface, create structural flaws and in turn bring about floccus mottling (refer to Chapter 6 Section 1 Floccus Mottle) (see pictures 6-1-90 and 6-1-91).

Due to the several different chemical components used in the restoration of ceramics, the characteristics of overlay mottles created by these materials can be widely varying. Using the sample pieces from Thailand, the materials used create a white, opaque film over the glaze surface (see picture 6-1-92); a small portion resembling needlepoint mottling. When reassembling pieces, trace amounts of dust and colored grains become trapped in said material (see picture 6-1-93). Some samples taken from Jingdezhen utilized clear glue (see picture 6-1-94).

It is very difficult to tell the difference between glue, natural overlay mottles, and needlepoint mottles, and there is a need for naked-eye inspection to tell whether or not an

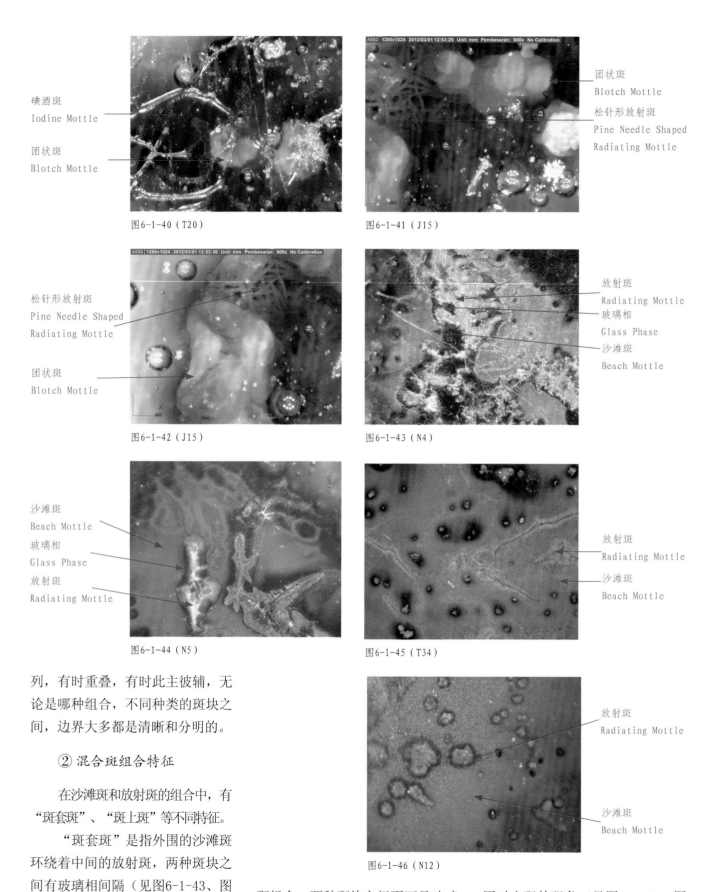

图6-1-40（T20）　图6-1-41（J15）
图6-1-42（J15）　图6-1-43（N4）
图6-1-44（N5）　图6-1-45（T34）
图6-1-46（N12）

列，有时重叠，有时此主彼辅，无论是哪种组合，不同种类的斑块之间，边界大多都是清晰和分明的。

② 混合斑组合特征

在沙滩斑和放射斑的组合中，有"斑套斑"、"斑上斑"等不同特征。

"斑套斑"是指外围的沙滩斑环绕着中间的放射斑，两种斑块之间有玻璃相间隔（见图6-1-43、图6-1-44）。

"斑上斑"是指放射斑与沙滩斑组合，两种斑块之间不可见玻璃相（见图6-1-45、图6-1-46）。

也有"斑套斑"、"斑上斑"同时出现的现象（见图6-1-47、图6-1-48）。

observed item has undergone repair.

### d. Bubble Root

#### i. Bubble Root Defined

A Bubble Root (mottle) is a depressed shape mottle with internal sediments, formed by the progression of a broken mouth bubble on the surface of YUB glaze under the influences of the post-firing environment.

#### ii. Formation of Bubble Roots

After YUB glaze undergoes corrosive weathering, a small portion of air bubbles within the glaze level become exposed, leaving a deep depressed pit in the glaze surface, also called a bubble root. Completely exposed to the environment, foreign materials are easily accumulated within the pit leaving deposited sediments on the internal wall, thus creating a bubble root mottle. Usually, it is the mouth portion of the ceramic piece that presents with bubble roots.

#### iii. Properties of Bubble Roots

The majority of bubble roots retain their inherent round shape, with a small portion also having ovular or irregular shapes due to the continual corrosive weathering to the bubble root. The most commonly seen colors in bubble roots are: Yellow, brown, and black. Some bubble roots can present with a spectrum of colors, with several different roots on the same piece presenting with different colors. Additionally, sedimentation can extend beyond the edge of the original bubble (see pictures 6-1-95 — 6-1-97).

### e. Floccus Mottle

#### i. Floccus Mottle Defined

Floccus Mottle is primarily white colored appearance within the YUB glaze level brought about by the post-firing environment; it has an irregular, free flowing, and disorderly shape that was formed by the depositing of soluble and partially soluble salts, dust particles, and other materials.

#### ii. Formation of Floccus Mottles

The fired ceramic glaze surface has randomly distributed cracks and openings. Additionally, as soil contains acidic and alkaline solutions that corrode the glaze surface, small cracks and openings in the glaze surface can be further created. It is through these small cracks and openings that foreign material can enter into the glaze level, forming soluble and partially soluble salts.

When the water in the soil evaporates, the soluble salts crystallize. However, due to the solubility of these salts, surface level water can dissolve and wash them away, Partially soluble salts can be washed away in the same manner. Therefore, the rate of floccus mottling is relatively low across the surface, but common within the glaze level.

In environments of alternating moisture, it is difficult for partially soluble salts to be dissolved away from the glaze level. Soluble salts dispersed within the glaze level crystallized after the evaporation of water and trap with them microscopic dust particles within the glaze level; these sedimentations form floccus mottles.

Soil and the glaze surface also contain positive and negative ions that, in reaction with each other, create partially soluble salt deposits of varying colors.

#### iii. Properties of Floccus Mottles

Floccus Mottles are primarily white in color, but can present with different colors (i.e. yellow) of partially soluble salt deposits. Within the glaze level, these mottles mainly assume irregular blotch and spot shapes, similar to wisps of clouds. Floccus mottles commonly occur around glaze flaws and breaks, for example:

Near crazing (see pictures 6-1-98 and 6-1-99),

Pitting (see pictures 6-1-100 and 6-1-101),

Overlay Mottles (see pictures 6-1-102 and 6-1-103),

Scrapes (see picture 6-1-104),

Other areas of imperfection (see picture 6-1-105).

It is also possible to see floccus mottling spreading into the glaze level around the flaw periphery.

On the samples from Thailand, we see a large amount of floccus mottling presenting, while pieces from within China rarely present with such. This is probably due to the moist environment found in Thailand, which has created a good setting for floccus mottles to form.

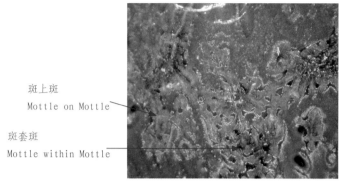

斑上斑 Mottle on Mottle
斑套斑 Mottle within Mottle

图6-1-47（N6）

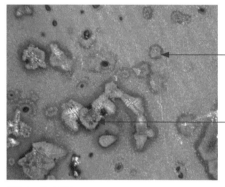

斑上斑 Mottle on Mottle
斑套斑 Mottle within Mottle

图6-1-48（H2）

（5）小结

元青花瓷器胎和釉中均含有铁元素，但青料中含铁量是最高的。不同种类、不同形貌的析晶斑，有的溢出釉表面，有的存在于釉层中，它们大都出现在青料浓处，大部分都含一定量的铁氧化物。

所有含铁氧化物的析晶斑，它们形貌和颜色的多样性，与釉的粘度及厚度有关，与分相物质的张力、密度、铁元素的价位都有关。

注：在现今采集的元青花标本的显微成像图片上，是已经风化的先天性斑块，我们能看到所有的先天性斑块特征都已是经过了800年左右时间的历练，即使保存再好的瓷器也会有一定程度的风化。我们无法回到元代探索元青花瓷器先天斑块的原貌，只能在迹型风化的绝对时间内，根据显微成像的图片情况，运用科学知识加以推理，想象元青花先天性斑块的本来面目。

## （二）后天性元青花斑块

### 1. 后天性元青花斑块概念

后天性元青花斑块是元青花瓷器在后天环境影响下形成的，以不规则形貌分布在釉表面及釉层中的各类斑块。

### 2. 后天性元青花斑块种类

（1）针眼斑

① 针眼斑概念

针眼斑是元青花瓷器在后天环境影响下，在釉表面以大小较均匀、深浅大致相同、疏密不定的聚集式独立微孔形式，呈不规则形貌随机分布的斑块。

通常情况下，在元青花析晶斑无釉覆盖处、釉表面结构有缺陷处，或曾有泥沙等异物覆盖处，较先以针眼斑形式开始风化。

② 针眼斑分类

针眼斑是随机分布的，它们可以在釉表面任何一个地方形成，形貌主要分为三种：

A. 雾状针眼斑

雾状针眼斑均匀细腻，呈雾状，边界处一般有由浓到淡的清晰过渡；存在于无釉覆盖的铁氧化物暴露处（见图6-1-49—图6-1-52）。

B. 磨砂针眼斑

磨砂针眼斑的针眼较雾状针眼更大，分布的密集程度及风化强度都大于雾状针眼斑，有磨砂感。磨砂针眼斑大致分为两种：

第一种磨砂针眼斑属于雾状针眼斑的发展，是风化强度更高的表现，分布密集，存在于无釉覆盖的铁氧化物暴露处（见图6-1-53—图6-1-55）。

第二种磨砂针眼斑的主要成因为，物盖物侵蚀釉表面，清除物盖物后暴露出磨砂针眼斑。为了区别，我们称这一类为物盖磨砂针眼斑。这类针眼斑分布随机，形状不规则，质地细腻，一般没有由浓到淡的过渡，但有清晰的界线，边界呈不规则曲线，正如不规则的海岸线（见图6-1-56、图6-1-57）。

C. 颗粒针眼斑

一般来说，颗粒针眼斑的针眼较磨砂针眼更大，针眼较前两类针眼斑更深，风化强度也更高。颗粒针眼斑的针眼呈颗粒状，针眼之间

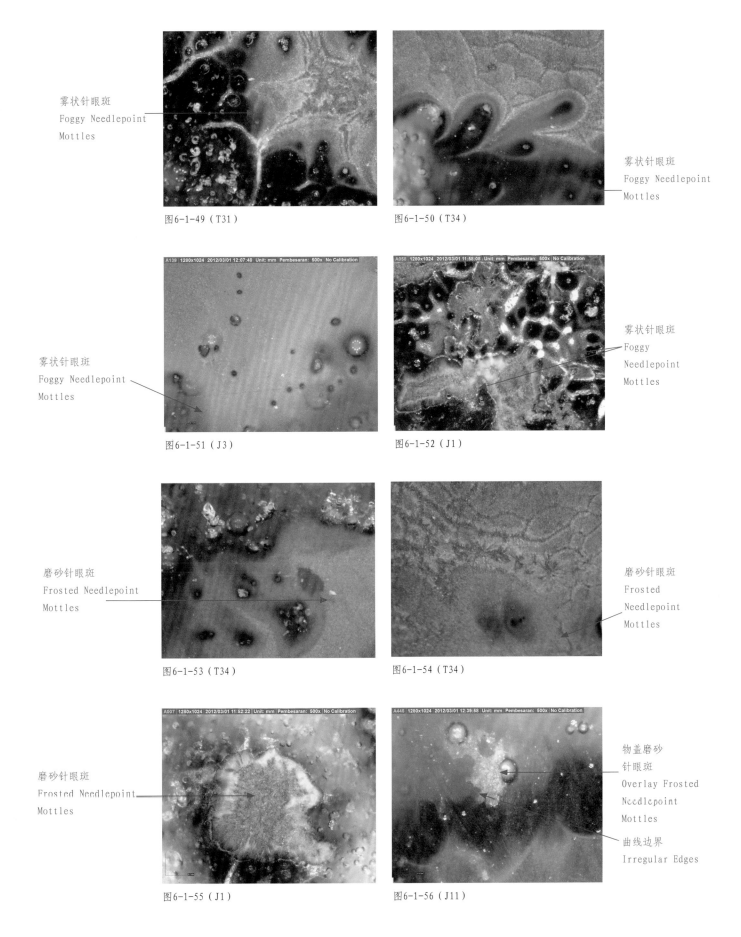

图6-1-49（T31） 雾状针眼斑 Foggy Needlepoint Mottles

图6-1-50（T34） 雾状针眼斑 Foggy Needlepoint Mottles

图6-1-51（J3） 雾状针眼斑 Foggy Needlepoint Mottles

图6-1-52（J1） 雾状针眼斑 Foggy Needlepoint Mottles

图6-1-53（T34） 磨砂针眼斑 Frosted Needlepoint Mottles

图6-1-54（T34） 磨砂针眼斑 Frosted Needlepoint Mottles

图6-1-55（J1） 磨砂针眼斑 Frosted Needlepoint Mottles

图6-1-56（J11） 物盖磨砂针眼斑 Overlay Frosted Needlepoint Mottles; 曲线边界 Irregular Edges

独立且界线分明，通常分布的密集程度不及雾状及磨砂针眼斑。颗粒针眼斑大致也分为两种：

第一种颗粒针眼斑主要分布在无釉覆盖的铁氧化物暴露处（见图6-1-58）或釉层结构缺陷处（见图6-1-59）。

第二种颗粒针眼斑的主要成因是：物盖物侵蚀釉表面，清除物盖物后，暴露出颗粒针眼斑。为了区别，我们称这一类为物盖颗粒针眼斑（见图6-1-60）。

③ 针眼斑分布状况

A. 针眼斑在流淌斑上的状况

a. 针眼斑在流淌斑上的形成

流淌斑主要成分为铁氧化物，铁氧化物在空气中容易风化。烧制成后，流淌斑上无釉层保护的部分最先风化，形成针眼斑。有釉层保护的，在后天釉层受到侵蚀后也以针眼斑形式开始风化。

b. 针眼斑在流淌斑上的特征

针眼斑反光呈现白色，覆盖在釉上流淌斑（沙滩斑）上，有的地方密集，有的地方稀疏。一般情况下，针眼斑风化的轨迹只在流淌斑形貌范围内。

有些雾状针眼斑均匀得像雾化的喷漆喷在釉表面，边沿处趋淡而均匀，尽管与玻璃相之间没有明显的边界线，但有一个从浓到淡的均匀过渡（见图6-1-61）。

有些磨砂针眼斑，通常密集地分布在沙滩斑的中心位置，因为那里釉层保护较差，风化强度相对较高。越往沙滩斑边沿靠近，受釉层保护越好，针眼斑越稀疏，过渡为雾状针眼斑（见图6-1-62）。

磨砂针眼斑一般没有浓淡过渡，但有边界感，这使沙滩斑富有层次。

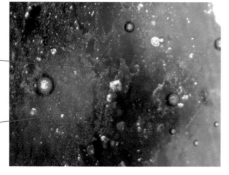

图6-1-57（B7）

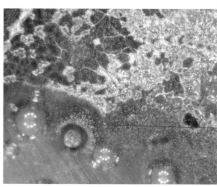

图6-1-58（H1）

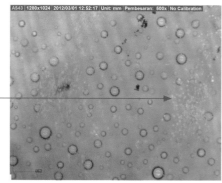

图6-1-59（J14）

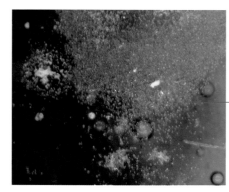

图6-1-60（T21）

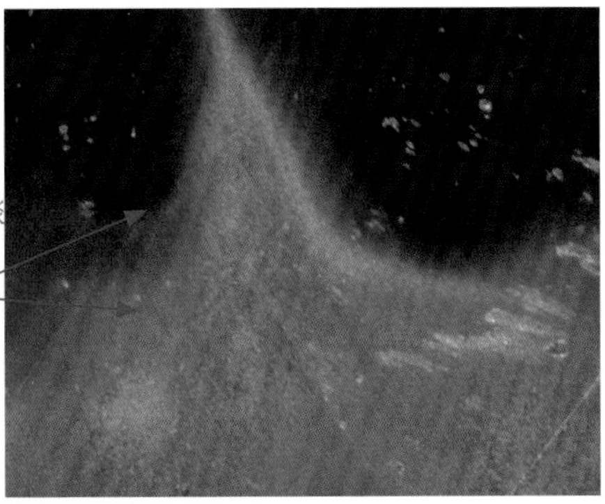
雾状针眼斑从浓到淡均匀过渡
Foggy Needlepoint Mottle Density Spectrum

图6-1-61（T21）

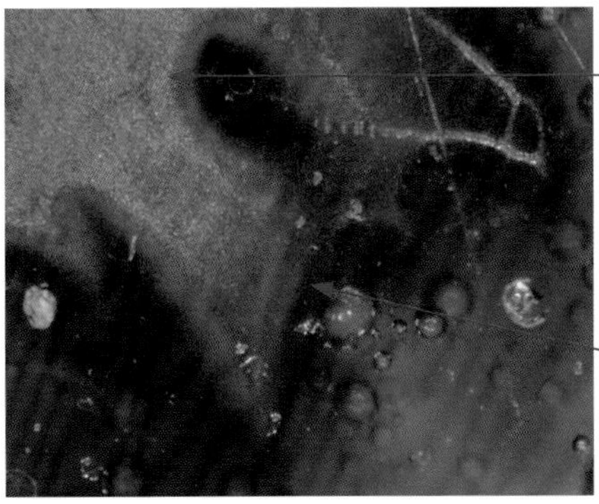
磨砂针眼斑
Grainy Needlepoint Mottles
雾状针眼斑
Foggy Needlepoint Mottles

图6-1-62（T13）

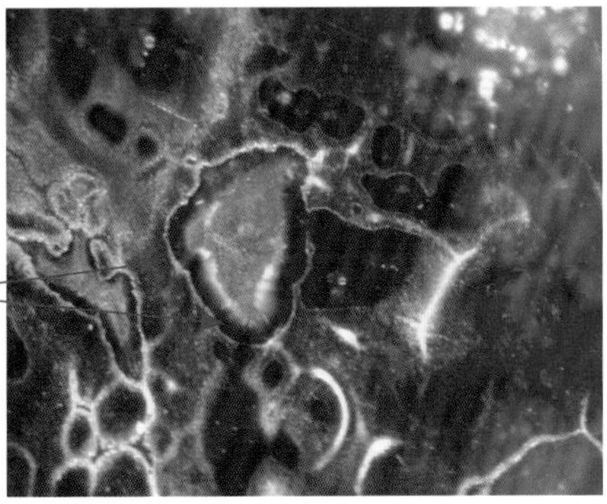
外实内虚
Inside Empty, Outside Full

图6-1-63（H4）

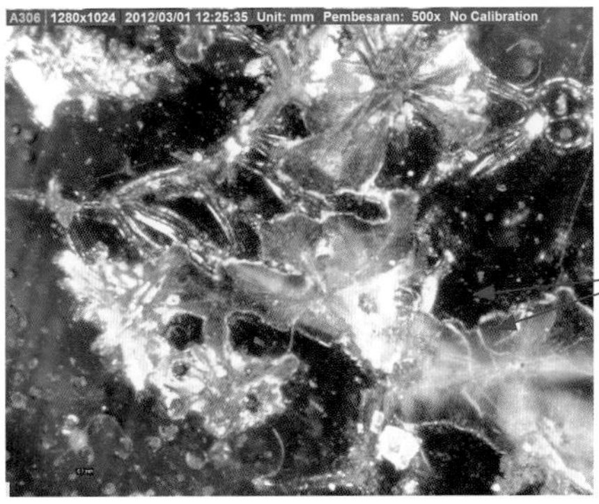
A306 1280x1024 2012/03/01 12:25:35 Unit: mm Pembesaran: 500x No Calibration
外虚内实
Inside Full, Outside Empty

图6-1-64（J6）

B. 针眼斑在放射斑上的状况

a. 针眼斑在放射斑上的形成

放射斑含一定量的铁氧化物，铁氧化物在空气中容易风化。烧制成后，无釉层保护的放射斑最先风化，形成针眼斑。有釉层保护的，在后天釉层受到侵蚀后也以针眼斑形式开始风化。

b. 针眼斑在放射斑上的特征

由于放射斑釉表形貌肌理较流淌斑更为复杂，所以，针眼斑在放射斑上的特征较流淌斑更复杂。

针眼斑反光呈现白色，分布在釉表面放射斑上，有的地方密集，有的地方稀疏。

有些雾状针眼斑分布在放射斑外围的玻璃相上，与放射斑界线分明，放射斑上的风化不明显，属于“外实内虚”现象（见图6-1-63）。

有些雾状针眼斑分布在放射斑界线内，与放射斑周围的玻璃相界线分明，而玻璃相的风化不明显，属于“外虚内实”现象。一般来说，雾状针眼斑出现在受到一定釉层保护的放射斑上，由于风化强度较低，针眼斑就像一层薄纱覆盖在放射斑上，甚至还能看清放射斑的原态（见图6-1-64）。

有些磨砂针眼斑顺着放射斑的析晶形状密集分布，通常分布在放射斑的中心位置，颜色泛白，一般不超出放射斑的范围，没有浓淡过渡（见图6-1-65、图6-1-66）。

部分放射斑外围会有一条边包裹，包边所在处属于两种不同相的分界处，是凹槽，结构有缺陷，容易风化。

包边的颜色在显微成像下有时呈白色，有时呈黄色。包边整体纤细、精致，少量的稍粗，但总体都是清晰的，走势大致随着放射斑的形状，但与斑块边缘之间留有一点空间（见图6-1-67）。

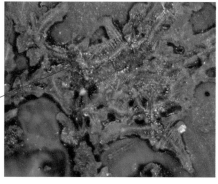

图6-1-65（N5） 图6-1-66（H5）

图6-1-67（N4） 图6-1-68（N3）

C. 针眼斑在团状斑上的分布状况

a. 针眼斑在团状斑上的形成

部分团状斑含一定量的铁氧化物，铁氧化物在空气中容易风化。大部分团状斑都处于釉层中，少有处于釉表面的情况。但在后天釉层受到侵蚀后，团状斑也以针眼斑形式开始风化。

b. 针眼斑在团状斑上的特征

由于大部分团状斑的铁氧化物含量较低，或者受釉层保护较好，风化强度较低，雾状针眼斑稀疏，像一层薄纱覆盖在团状斑上，甚至还能看清团状斑的原态（见图6-1-68）。

与部分放射斑类似，部分团状斑外围也会有一条边包裹，包边总体也是精致、收敛、清晰的，走势大致随着团状斑的形状，但与斑块边缘之间留有一点空间（见图6-1-68）。

D. 针眼斑在玻璃相上的分布状况

a. 针眼斑在玻璃相上的形成

针眼斑在玻璃相上的形成大致有两种原因：

一是釉表面结构缺陷处如分相结合处易遭风化形成针眼斑。

二是釉表面受到外界物质长期覆盖后，清除掉物盖物，会在物盖处的釉表面留下针眼斑，这类针眼斑是长期物盖侵蚀所致。

b. 针眼斑在玻璃相上的特征

一般针眼斑都反光呈白色。

颗粒针眼斑：在釉表面结构缺陷处（见图6-1-69）；在无釉覆盖的铁氧化物暴露处（见图6-1-70）。

物盖磨砂针眼斑：与周围未受到侵蚀的玻璃相界线分明（见图6-1-71）。

物盖颗粒针眼斑：针眼分布独立，针眼之间界线分明，与周围未受到侵蚀的玻璃相界线分明（见图6-1-72）。

（2）坑点斑

① 坑点斑概念

坑点斑是元青花瓷器在后天环境影响下，在釉表面形成的，深浅不一、大小不一、形貌不一的不规则凹形斑块。

② 坑点斑形成

坑点斑的形成原因主要有四种：

A. 由C形、O形纹发展而成：

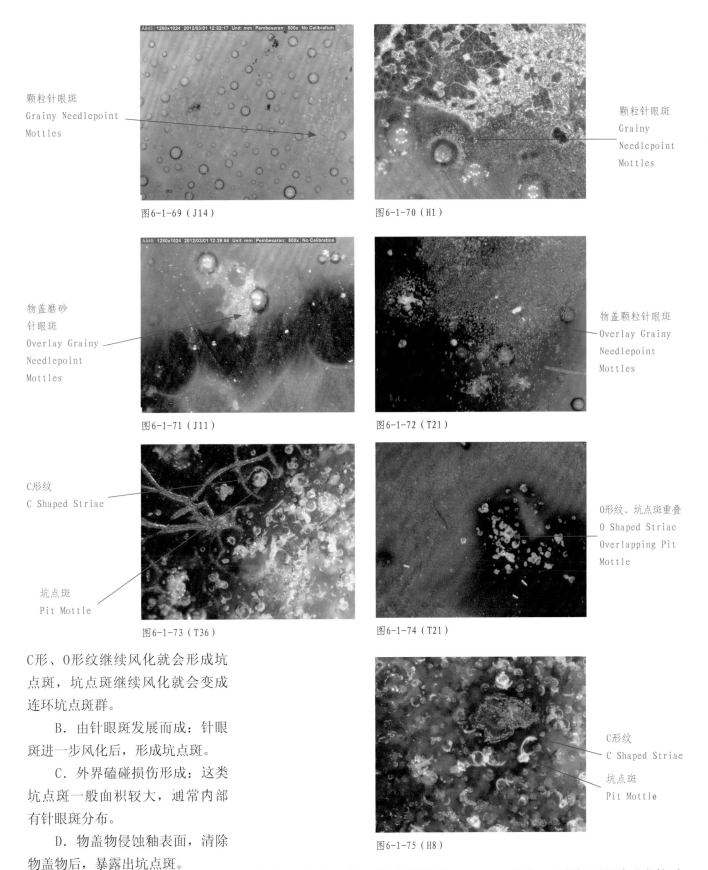

图6-1-69 (J14) — 颗粒针眼斑 Grainy Needlepoint Mottles
图6-1-70 (H1) — 颗粒针眼斑 Grainy Needlepoint Mottles
图6-1-71 (J11) — 物盖磨砂针眼斑 Overlay Grainy Needlepoint Mottles
图6-1-72 (T21) — 物盖颗粒针眼斑 Overlay Grainy Needlepoint Mottles
图6-1-73 (T36) — C形纹 C Shaped Striae；坑点斑 Pit Mottle
图6-1-74 (T21) — O形纹、坑点斑重叠 O Shaped Striae Overlapping Pit Mottle
图6-1-75 (H8) — C形纹 C Shaped Striae；坑点斑 Pit Mottle

C形、O形纹继续风化就会形成坑点斑，坑点斑继续风化就会变成连环坑点斑群。

B. 由针眼斑发展而成：针眼斑进一步风化后，形成坑点斑。

C. 外界磕碰损伤形成：这类坑点斑一般面积较大，通常内部有针眼斑分布。

D. 物盖物侵蚀釉表面，清除物盖物后，暴露出坑点斑。

③ 坑点斑特征

由C形、O形纹发展而成的坑点斑，有时在坑点斑上还能见C形、O形纹路，形成C形、O形纹与坑点斑共存的现象。这是最常见的一种坑点斑（见图6-1-73—图6-1-75），反映了迹型形成的时间过程性。

一开始坑点斑是独立零星地分布的，如果风化继续，坑点斑

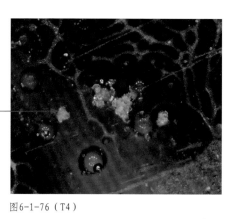

图6-1-76（T4） 连环坑点斑群 Pit Group

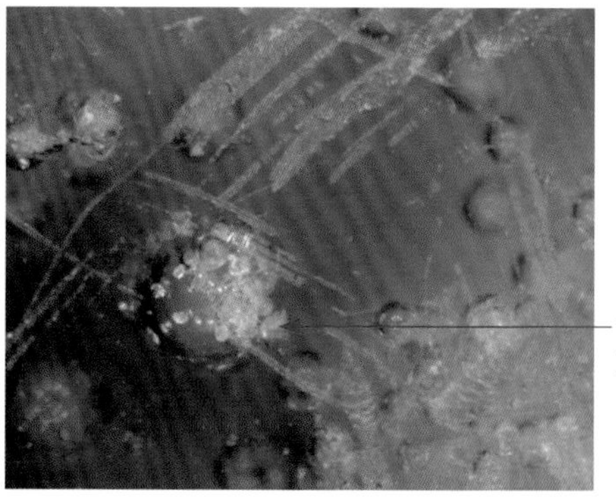

图6-1-77（T12） 连环坑点斑群 Pit Group

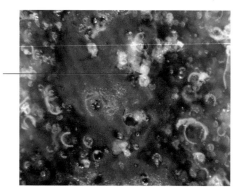

图6-1-78（H8） 连环坑点斑群 Pit Group

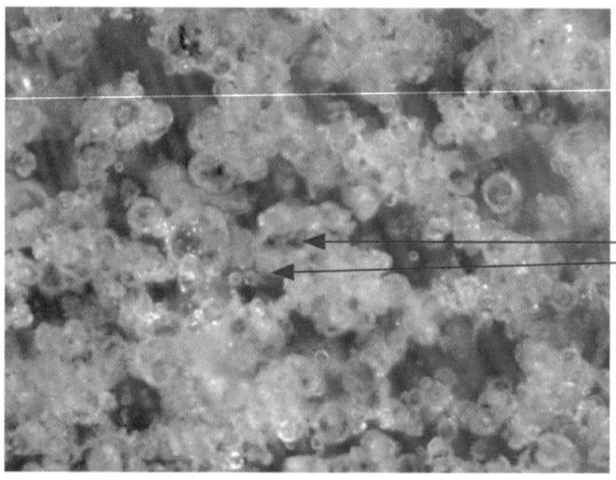

图6-1-79（B1） 坑点斑有色阶变化 Pit Mottle with Color Spectrum

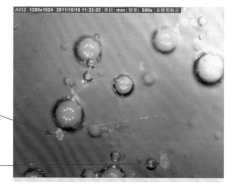

图6-1-80（DY1） 细边 Fine Border；针眼斑 Needlepoint Mottles

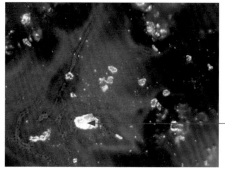

图6-1-81（B3） 部分针眼斑不可见 Needlepoint Mottles Partially Invisible

之间相连贯通，就变成重叠连环状、有螺旋感的坑点斑群，通常呈不规则形状（见图6-1-76—图6-1-78）。

由于其凹形的特征，坑点斑内易积聚有色物质，有的坑点斑部分呈黄色，有色阶变化（见图6-1-79）。

由针眼斑发展而成的坑点斑分布随机，一般呈白色，大小不一。有的坑点斑被曲折的细边包裹，斑内还可见针眼斑分布（见图6-1-80）；有的坑点斑内针眼斑分布密集，密集处已看不见独立的针眼斑，且不透影（见图6-1-81）。

针眼斑与由针眼斑发展而成的坑点斑的区别是：

坑点斑无论大小均各自独立，形貌各异，深浅不一。坑点斑之间能观察到底板，坑点斑与底板的界线是清晰的（见图6-1-82）。

而大部分针眼斑，针眼与针眼之间几乎观察不到底板，就像种稻子播种秧田的要求："泥不见天，谷不重叠"。虽然有些针眼斑分布也稀疏，但其大小、深浅还是均匀的（见图6-1-83）。

因外界磕碰损伤形成的坑点斑面积较大，还能够观察到斑块内的针眼斑和沉积物，常有色阶变化（见图6-1-84）。

瓷器长期存放于干湿交替的环境中，在坑点斑下的釉层中可形成絮状斑（参见本章本节絮状斑部分）（图6-1-85）。

不同成因的坑点斑可以共存（见图6-1-86）。

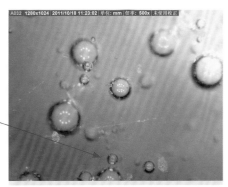
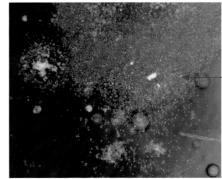

坑点斑之间的玻璃相清晰可见
Glass Phase Clearly Visible Amongst Pit Mottles

针眼斑之间几乎不见玻璃相
Glass Phase Almost Inobservable Amongst Needlepoint Mottle

图6-1-82（DY1）　　　　　图6-1-83（T21）

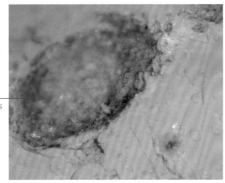
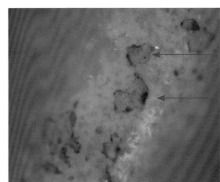

坑点斑内沉积物、针眼斑，有色阶变化
Sediments, Needlepoint Mottles and Color Spectrum within Pit Mottle

坑点斑
Pit Mottle

絮状斑
Floccus Mottles

图6-1-84（T5）　　　　　图6-1-85（T31）

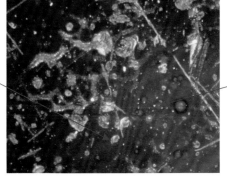

C形、O形纹发展成坑点斑
Progression of C and O Shaped Striae into Pit Mottle

针眼斑发展成坑点斑
Progression of Needlepoint Mottle into Pit Mottle

图6-1-86（T7）

（3）物盖斑

① 物盖斑概念

物盖斑是元青花瓷器在后天环境影响下，釉表面被外界物质长期覆盖而形成的斑块。

② 物盖斑形成

元青花瓷器上的物盖斑大都是后天环境中，瓷器周围的泥沙等物质长期覆盖在瓷器表面形成的，也有后来人们对瓷器修补时留下的黏合剂印记。

③ 物盖斑特征

物盖斑颜色主要有黄色、褐色及灰白色，斑块与周围未被覆盖的底板界线是分明的，不像针眼斑的边界多是曲线，其边界有直线、有曲线。

斑块有层次，有的呈起皮状（见图6-1-87），有的有龟裂纹，像旱地（见图6-1-88），有的物盖斑上有划痕（见图6-1-89）。

有的物盖斑长时间侵蚀釉表面，造成釉表面结构破损，易形成絮状斑（参见本章第一节絮状斑部分）（见图6-1-90、图6-1-91）。

修补瓷器留下的印记根据其使用的化学用剂的不同，其物盖斑形貌也不同。如泰国标本的黏合剂，剂量多处呈白色，不透明（见图6-1-92），剂量少处形貌如针眼斑，掺杂着灰尘及其他有色颗粒（见图6-1-93）；部分景德镇标本的黏合剂呈透明状（见图6-1-94）。

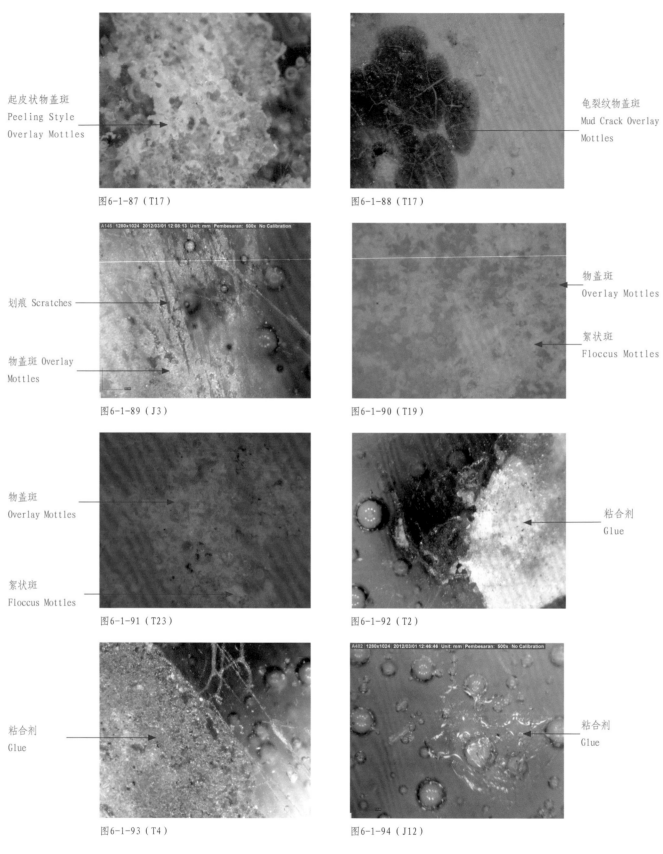

图6-1-87（T17） 图6-1-88（T17）
图6-1-89（J3） 图6-1-90（T19）
图6-1-91（T23） 图6-1-92（T2）
图6-1-93（T4） 图6-1-94（J12）

有时黏合剂与物盖斑、针眼斑形貌相近，区别难分，必要时需要结合肉眼来观察瓷器。

（4）泡根斑

① 泡根斑概念

泡根斑是元青花瓷器在后天环境影响下，由破口泡发展形成，内有沉积物的凹形斑块。

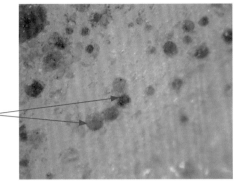

图6-1-95（T6）　　　　图6-1-96（T2）

有色差变化泡根斑
Color Disparity Bubble Roots

不规则形状泡根斑
Irregular Shaped Bubble Root

不规则形状泡根斑，多种沉积物有色阶变化
Irregular Shaped Bubble Root, Many Types of Sediments and Color Spectrum within Bubble Root

图6-1-97（H7）

② 泡根斑形成

元青花瓷器的釉表面被风化侵蚀或磨损后，存在于釉层中的部分气泡无釉层包裹，完全暴露在外界环境中，变成泡根，形成深浅不一的凹坑，外界物质长期在泡根内沉积形成泡根斑。一般来说，陶瓷的口沿处易形成泡根斑。

③ 泡根斑特征

泡根斑大部分是圆形，也有椭圆形和不规则形，这是因为泡根边缘继续风化被侵蚀，所以形成不规则形状。泡根斑颜色有黄色、褐色及黑色等。有的同一斑块内有多种沉积物，且有色阶变化，有的临近的几个斑块有色差（见图6-1-95—图6-1-97）。

（5）絮状斑

① 絮状斑概念

絮状斑是元青花瓷器在后天环境影响下，在釉表面及釉层中，由可溶盐、难溶盐和灰尘等物质沉积形成的以白色为主，呈无序状，具有流淌、渗透、漫延趋势的不规则斑块。

② 絮状斑形成

陶瓷釉表面随机分布有裂纹或孔隙。此外，土壤中具有一定酸或碱性的水溶液会侵蚀元青花瓷器釉表面，使釉表面形成裂纹或孔隙。外界物质通过这些裂纹和孔隙进入釉层，在釉层中形成可溶盐及难溶盐。

当土壤中水分蒸发，可溶盐在釉表面及釉层中析出。在釉表面的可溶盐易因溶于水而流失，难溶盐也易掉落，所以絮状斑在釉表面存在的几率较小，普遍存在于釉层中。

在干湿交替的环境中，渗入釉层中的难溶盐难以流失，可溶盐会在水分蒸发后的釉层中形成结晶，它们连同一些细小灰尘在釉层中叠加沉积形成絮状斑。

土壤中的及釉表面的阴阳离子会发生反应，形成不同颜色的难溶盐沉积物。

③ 絮状斑特征

絮状斑以白色为主，但也有少量其他不同颜色的难溶盐沉积物（比如黄色），一般都以块状、片状等不规则形貌分布在釉层中，形貌如絮状白云。絮状斑一般形成在釉层缺陷处或破损处，如裂纹下方（见图6-1-98、图6-1-99）、坑点斑处（见图6-1-100、图6-1-101）、物盖斑处（见图6-1-102、图

6-1-103)、划痕处（见图6-1-104）和其他釉损处（见图6-1-105）等。

也有的絮状斑会扩散到釉损处周边的釉下。

在泰国出土的标本上看到大量絮状斑的存在，而国内的标本则少见。这应该是由于泰国的环境较潮湿，容易形成絮状斑。

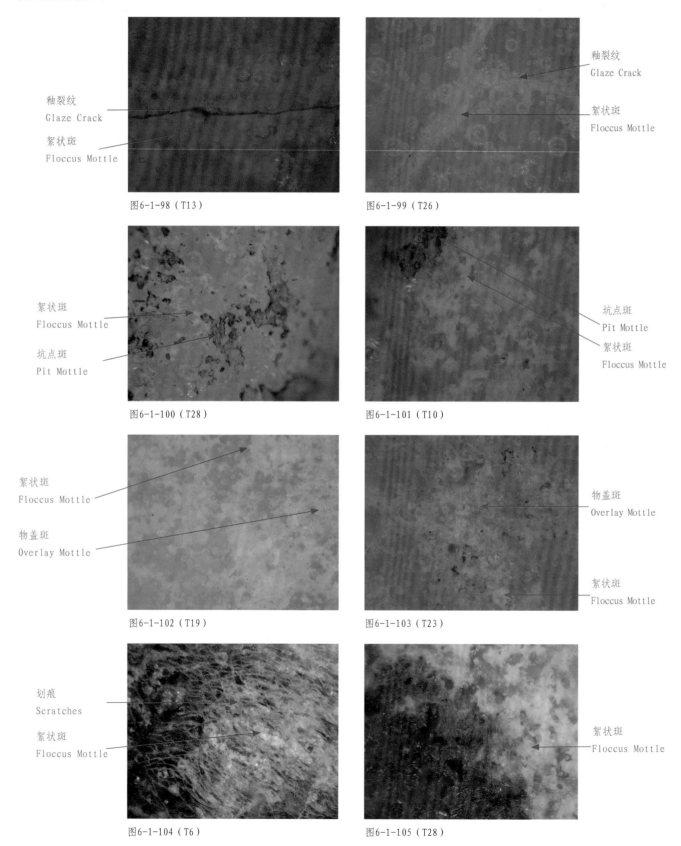

图6-1-98（T13）　　　　　　　图6-1-99（T26）

图6-1-100（T28）　　　　　　图6-1-101（T10）

图6-1-102（T19）　　　　　　图6-1-103（T23）

图6-1-104（T6）　　　　　　　图6-1-105（T28）

# 第二节
# 元青花纹路类别

## 一、元青花纹路概念

元青花纹路是元青花瓷器在烧制过程中和后天环境影响下形成的，以线状、条状、网状等形貌分布在釉表面和釉层中的迹型类别。

## 二、元青花纹路种类

### （一）流淌纹

#### 1. 流淌纹概念

流淌纹是元青花瓷器上以铁为主，在烧制过程中和后天环境影响下形成的，以线状、条状、网状等形貌，分布在釉表面和釉层中的纹路。

#### 2. 流淌纹形成

元青花瓷器在高温烧制时，青料及釉料会逐渐熔化成熔融态，因浮力及气泡的搅动，胎、釉料以及青料中的铁氧化物在表面张力的作用下，在釉层中集聚，也可再形成分相。

铁氧化物易在气泡边缘聚集，部分会随气泡升至釉表面，顺着凹凸不平的釉表面向四周扩散，流动铺展，这就是一个润湿过程。

当温度下降时，这些铁氧化物便在釉表面析出，铁氧化物数量多的形成斑块，铁氧化物少的形成纹路，就是流淌纹。

位于釉表面铁氧化物含量较高的流淌纹易风化形成凹槽，就是阴纹；$SiO_2$含量较高的流淌纹不易被风化，相对凸起，形成阳纹。

釉层中未升至釉表面的铁离子在高温下熔融在釉料中，釉层冷却凝固时在釉层中形成流淌纹。

#### 3. 流淌纹分类

流淌纹可分为三类：碘酒纹、阴纹、阳纹。

##### （1）碘酒纹

釉层中受釉层保护，风化不明显的流淌纹为碘酒纹。碘酒纹颜色呈褐色，因与液体碘酒相似而得名。纹路呈溪流流淌状，颜色有深有浅（见图6-2-1）。

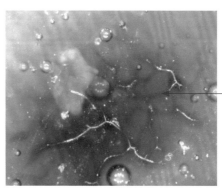

图6-2-1（N2）

碘酒纹
Iodine Striae

##### （2）阴纹

阴纹可分成中空型阴纹、针眼型阴纹及细线型阴纹。颜色主要有银白色或橘黄色。

① 中空型阴纹

中空型阴纹有两条泛银白光的细边，细边内的部分呈玻璃相，这应该是由于中空型阴纹所在的凹槽底部还有釉层保护，风化作用不明显，所以纹路内部像是"中空"的。在同一图片上，同一宽度的纹路内有时会有多条反光白边，边与边有的间断，有的错综（见图6-2-2）。

② 针眼型阴纹

针眼型阴纹是中空型阴纹进一步风化的结果。针眼型阴纹两条细边内的部分被针眼斑不均匀地覆

# Section 2:
# The Yuan Dynasty Underglaze Blue Porcelain Striae Category

## 1. YUB Striae Defined

Yuan Underglaze Blue Striae are the category of line, strip, and web shaped appearances, created during the firing process and acquired under the influences of the post-firing environment, that are disbursed across the glaze surface and within the glaze level of Yuan Dynasty Underglaze Blue Porcelain.

## 2. YUB Striae Types

### A. Flowing Striae

#### I. Flowing Striae Defined

Flowing Striae are line, strip, and web shaped appearances across the surface and within the glaze level of YUB, created during the firing process or acquired under the influences of the post-firing environment, that consist primarily of Iron.

#### II. Formation of Flowing Striae

During the firing of YUB, cobalt-blue dye and glaze gradually transitions into a molten state. Due to the buoyancy and movement of the air bubbles within the glaze level, the Iron Oxides of the body, glaze, and dye under the effect of surface tension, accumulate together and can form into spilt phases.

It is easy for these Iron Oxides to accumulate around air bubbles; a portion is then carried to and dispersed throughout the uneven surface of the glaze. This flowing dispersion is a fluid process.

When the temperature decreases, a portion of the Iron Oxides begins to crystallize. Areas of high Iron Oxide form flowing mottles and areas of lower Iron Oxide form striae, which is the aforementioned flowing striae.

Areas of high Iron Oxide are easily corroded by weathering and leave slight depressions, or depressed striae; while areas of higher $SiO_2$ are resistant to weathering and remain elevated, forming raised striae.

Fe ions within the glaze, under high firing temperatures, melt into the glaze, and form iodine striae within the glaze level.

#### III. Types of Flowing Striae

There are three types of Flowing Striae: Iodine Striae, Depressed Striae and Raised Striae.

##### a. Iodine Striae

Under the protective covering of the glaze, these are flowing striae with indistinct weathering. Iodine striae consist of a brownish color similar to that of liquid iodine, hence the name. These striae assume the shape of flowing streams with varying degrees of color density (see picture 6-2-1).

##### b. Depressed Striae

Empty, Needlepoint, and Thin Striae are all types of Depressed Striae, and consist primarily of reflective or tangerine coloration.

###### i. Empty Striae

Empty Striae have two thin reflective outlines, with the interior appearing as glass phase. This is due to the fact that the bottom most portion of the groove retained a protective glaze layer; thus making it less susceptible to weathering, and creating an 'empty' center. Within one picture we can see that

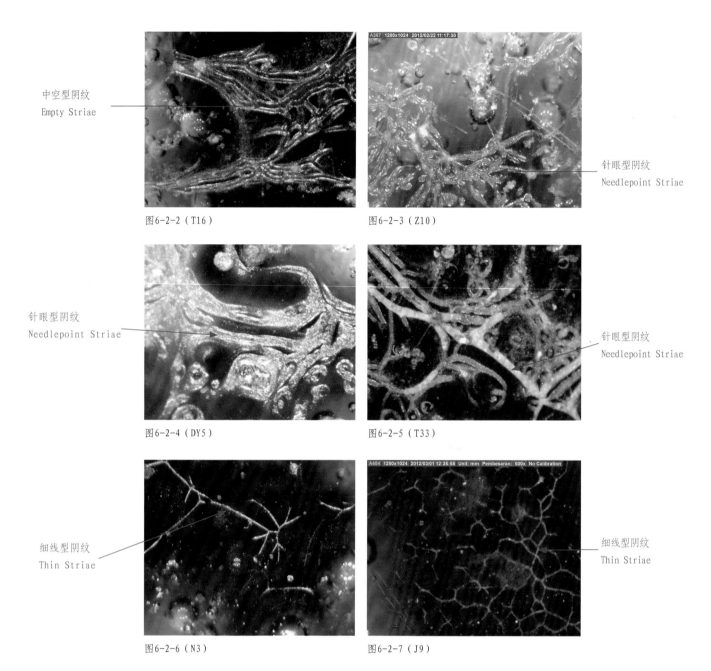

图6-2-2（T16） 图6-2-3（Z10）

图6-2-4（DY5） 图6-2-5（T33）

图6-2-6（N3） 图6-2-7（J9）

盖，针眼斑大部分呈银白色，少数呈橘黄色；无针眼斑覆盖处还是玻璃相。当风化作用加深后，凹槽底部逐渐被针眼斑覆盖，两条细边也逐渐观察不到了（见图6-2-3—图6-2-5）。

③ 细线型阴纹

细线型阴纹是细线状的，颜色大多呈银白色。这类纹路大多出现在釉表面风化强度较低，或釉层较厚，或铁氧化物数量较少的流淌斑上（见图6-2-6、图6-2-7）。

（3）阳纹

阳纹颜色大多呈褐色，它们基本都出现在沙滩斑上，大多数位置高于沙滩斑，显微成像为凸起部分，纹路饱满均匀（见图6-2-8、图6-2-9）。

4. 小结

流淌纹可呈现粗、细、深、浅的形貌变化，也有颜色的变化。同一张图片上可有上述多种流淌纹并存，它们的形貌、颜色及分布都有一定的规律。

流淌纹的形貌、颜色及其分布规律与流淌纹所处区域釉的含量、铁氧化物的数量以及釉表面的凹凸情况有关。

碘酒纹基本未风化，呈先天性原貌。

在碘酒斑上出现的流淌纹，尽管该区域釉表面存在凹槽，但凹槽处被釉表面保护得较好，$SiO_2$的

empty striae can present with several lines within a single striation, some being interrupted or intercrossing with other striae (see picture 6-2-2).

### ii. Needlepoint Striae

Needlepoint Striae are a weathering progression from empty striae. The internal section of the empty striae becomes covered by an uneven disbursement of needlepoint mottling and assumes a reflective or tangerine coloration. Areas that are untouched by the needlepoint mottling retain their glass phase characteristics. As most portions of the striae bottom become more weathered and covered by the needlepoint mottles, the thin outlines become gradually less evident (see pictures 6-2-3—6-2-5).

### iii. Thin Striae

Thin Striae present merely as very fine lines, primarily assuming reflective coloration; and occur in areas of very low degrees of glaze weathering, areas with exceptionally thick glaze, or flowing mottles that have a relatively low amount of surface Iron Oxide (see pictures 6-2-6 and 6-2-7).

### c. Raised Striae

Raised Striae assume a primarily brownish coloration. They mostly appear within beach mottles. In general, they are microscopically elevated lines that are full and evenly distributed across the mottle (see pictures 6-2-8 and 6-2-9).

### IV. Summary

Flowing striae can assume a wide variety of thickness, depth and coloration. A single microscopic image can present with several types of striae in conjunction with one another; with their appearance, color, and disbursement all differing.

The appearance, color, and disbursement of striae depends on the amount of glaze, amount of Iron Oxide, and topography of the glaze surface.

Iodine Striae are essentially unweathered, and thus retain their initial appearance.

On iodine mottles, due to the grooves in the glaze and higher glaze to Iron Oxide ratio, the most common types are thin and empty flowing striae; this is because of the lower weathering severity in that area. They usually assume a reflective nature.

As the glaze gradually becomes more weathered, flowing striae slowly transition into the needlepoint variety and assume a reflective or tangerine color.

In areas of beach mottling with high degrees of weathering, the areas of different composition to that of the beach mottle, specifically containing less Iron Dioxide, have lower weathering severity, thus forming thick raised striae of a brown coloration (see pictures 6-2-10 and 6-2-11).

Pictures 6-2-12—6-2-14 illustrate the three-dimensional topography of raised and depressed striae (picture 6-2-12 is the two-dimensional and pictures 6-2-13 and 6-2-14 are the three-dimensional view of the same spot on the same piece).

## B. C and O Shaped Striae

### I. C and O Shaped Striae Defined

C and O shaped lines disbursed across the surface of YUB porcelain glaze, in areas of structurally weakened glaze, acquired under the influence of the post-firing environment.

### II. Formation of C and O Shaped Striae

During the firing of YUB porcelain, split phases can form during the cooling of the glaze, some of them form a circular trace on the glaze surface and within the glaze level. In general, the stability of split phases is not as good as that of glass phase, and the area where two different phases meet has structural flaw. Therefore, they are most susceptible to weathering. Additionally, during the firing of YUB porcelain, air bubbles within the liquid glaze continually raise towards the surface, with some bursting out. When the glaze cools to a solid, the air pockets that burst through the glaze sometimes leave a circular trace on the glaze surface. Foreign materials are easily deposited within the grooves, present across the glaze face, and these areas can weather to become C and O shaped striae.

阳纹
Raised Striae

图6-2-8（T3）

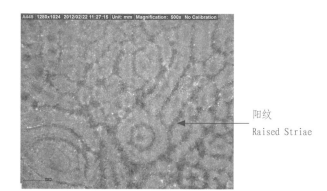

图6-2-9（Z12）

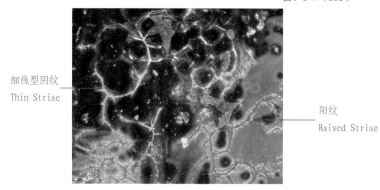

细线型阴纹
Thin Striae

阳纹
Raised Striae

图6-2-10（N2）

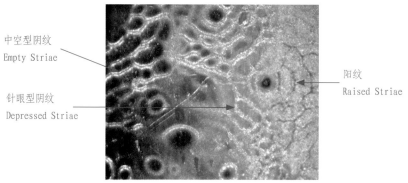

中空型阴纹
Empty Striae

针眼型阴纹
Depressed Striae

阳纹
Raised Striae

图6-2-11（DY3）

含量较高，因此风化强度较低，流淌纹主要以细线型及中空型阴纹出现，颜色主要呈银白色。

当釉表面逐渐被侵蚀，风化作用加深，流淌纹主要以针眼型阴纹出现，颜色主要呈银白色或橘黄色。

当逐渐进入几乎无釉覆盖的沙滩斑区域时，那里分布的流淌纹由于成分有别于沙滩斑，含铁氧化物较少，所以风化强度较沙滩斑轻，主要以阳纹出现，颜色呈褐色，显微成像为凸起部分（见图6-2-10、图6-2-11）。

图6-2-12—图6-2-14分别展现了

### III. Properties of C and O Shaped Striae

C shaped striae are semi-circle lines, and O shaped striae are circular and ovular lines that form visible appearances according to their shape. Their colors include yellow, brown, and black (see pictures 6-2-15—6-2-19).

C and O shaped striae all have an evident depth, extending into the glaze level. According to varying degrees of weathering, the thickness of these striae also varies, but they remain distinctly delineated from the surrounding glass phase.

C and O shaped striae, if further weathered, can form pit mottles; and it is common to see pitting in conjunction with these striae (see pictures 6-2-17 and 6-2-18).

In areas of extensive glaze weathering, it is possible for C and O shaped striae to present with floccus mottling as well (see picture 6-2-19).

## C. Glaze Crack

### I. Glaze Crack Defined

Also called crazing, Glaze Cracks are any appearance of straight, curved, or irregular lines splitting the glaze surface of YUB porcelain that was caused either during firing or under the influence of the post-firing environment.

### II. Formation of Glaze Cracks

Glaze cracks can be caused by any number of the following reasons:

a. During the firing process, the components of the glaze volatilize and create empty spaces. Due to the stress caused by the accumulation of these spaces into gaps, the glaze cracks.

b. If the glaze and the micro phases within the glaze expand at different rates, cracks in the glaze can occur.

c. The stress tension caused by the $R^+$ and $H^+$ ion exchange with $\equiv Si—O—R$, because of the $H^+$ ion being relatively smaller than the $R^+$ ion, can create glaze cracks [102].

d. The composition of glaze and clay body being different, their expansion rates are also different. As the porcelain cools, stress is created, which the thicker body can absorb, but the glaze reaches a breaking point at which it cracks.

e. Glaze lacks malleability. As such, when hard external forces collide with the glaze, the inability to absorb the blow can result in cracking.

Cracks are not necessarily a direct result of weathering; some are caused inherently during the firing process but some are indeed acquired under the influence of the post-firing environment. Additionally, not all cracks are visible under microscopic observation. If, after firing, the glaze encounters drastic temperature changes or the stress imposed by the crystallization of soluble salt or partially soluble salt deposits, the glaze cracks can become broadened.

There is a type of glaze crazing that is intentionally man-made, in the pursuit of aesthetic beauty, by utilizing the differences in body-glaze expansion rates.

However, the crazing that presents on YUB was not intentionally fired with this sort of artistic outcome in mind. Therefore, most crazing that presents is considered to be technical flaw.

### III. Properties of Glaze Cracks

Glaze cracks can either be straight or rounded in appearance.

Varying degrees of width is not always evident on a single crack (see pictures 6-2-20 and 6-2-21);

While some cracks can present with widely varying widths (see picture 6-2-22);

Some cracks have indistinct width and color changes (see picture 6-2-23);

While others are widely varying (see pictures 6-2-24 and 6-2-25).

In general, cracks that were created earlier are thicker and have a wider color spectrum than that of cracks that were formed later.

Some cracks present with several cracks branching out from it, similar to a root network (see picture 6-2-26).

Some cracks have entirely changed color, but have a limited color variation (see picture 6-2-20), while others still assume several colors along the spectrum of white, yellow, brown, and black (see pictures 6-2-21, 6-2-24 and 6-2-25).

Some cracks only undergo partial coloration (see picture 6-2-25).

阴纹和阳纹的三维形貌（图6-2-12—图6-2-14为同一器物同一位置图片，图6-2-12为二维图片，图6-2-13、图6-2-14为三维图片）。

### （二）C形纹、O形纹

#### 1. C形纹、O形纹概念

C形纹、O形纹是元青花瓷器在后天环境影响下，以字母C、O形貌分布在釉表面结构有缺陷处的线形纹路。

#### 2. C形纹、O形纹形成

在烧制元青花瓷器时，熔融的釉在冷却过程中又产生分相，有的以环状形式存在于釉表面及釉层中。一般情况下，分相物质的化学稳定性不及玻璃相，而且两相结合处结构有缺陷，易风化，从而逐渐形成C形纹、O形纹。另一方面，高温烧制元青花瓷器时，产生的气泡在熔融的釉料中不断上升，有的冲出釉表面。当釉料冷却凝固下来时，气泡冲出釉表面时留下的环状凹槽便会留在釉表面上，外界物质容易在釉表面的这些环状凹槽处沉积，逐渐风化形成C形纹、O形纹。

#### 3. C形纹、O形纹特征

C形纹是半圆形貌的纹路，O形纹是圆形、椭圆形貌的纹路。它们的颜色有黄色、褐色和黑色等（见图6-2-15—图6-2-19）。

C形、O形纹均进入釉层一定的深度，纹路风化强度不同，粗细也不一样，但都与玻璃相保持边界清

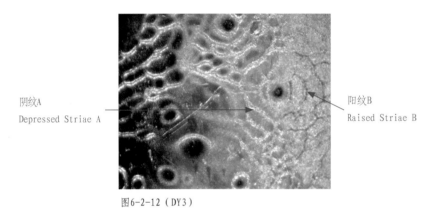

图6-2-12（DY3）

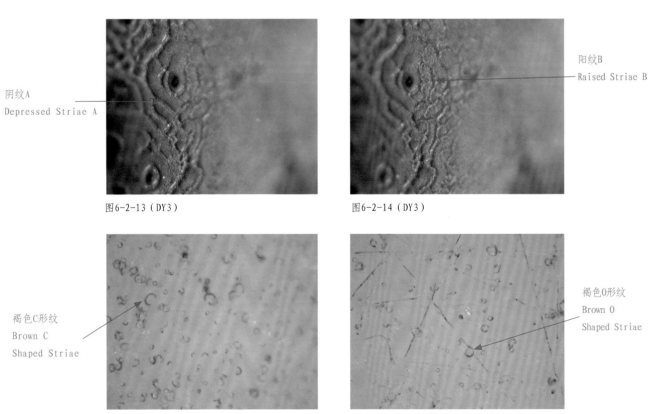

图6-2-13（DY3）　　　　　图6-2-14（DY3）

图6-2-15（S3）　　　　　图6-2-16（T4）

The characteristics of the areas surrounding cracks include:

Color or texture changed air bubbles (see pictures 6-2-27 and 6-2-28),

Underglaze color change (see pictures 6-2-29 and 6-2-30),

Underglaze floccus mottling (see pictures 6-2-26 and 6-2-31).

These characteristics can occur independently or concurrently.

## D. Scratches

### I. Scratches Defined

Scratches are the imperfections on YUB porcelain glaze that present as lines on the glaze surface that were left by the abrasive force of a solid external object.

### II. Formation of Scratches

During post-firing use, and due to coming into contact with hard objects in the surrounding environment, the glaze surface can become scratched.

### III. Properties of Scratches

The range of scratch characteristics include: Thick and thin, long and short, deep and shallow, straight and curved, with varying colors including white, yellow, brown, and black. They are random and unevenly dispersed across the glaze surface (see pictures 6-2-32 — 6-2-42).

Some scratches have 'ox hair' features; short and thin with wild, directionless distribution (see pictures 6-2-35 and 6-2-36).

At times, a single picture can present with varying appearances and colorations of scratches (see pictures 6-2-35, 6-2-37 and 6-2-38).

At places of deepening scratches, to the point of grooves within the glaze, it is easy for foreign objects to become lodged and corrode the glaze, forming needlepoint mottling (see pictures 6-2-38 and 6-2-39).

At times, deeps scratches can destroy the structural integrity of the surrounding glaze, in which case it is easy for floccus mottling to occur (see pictures 6-2-40 — 6-2-42).

晰。

有的C形、O形纹继续风化就会形成坑点斑。有时在C形、O形纹路下还能看见坑点斑，形成C形、O形纹与坑点斑共存的现象（见图6-2-17、图6-2-18）。

在釉表面风化强度较高处，C形纹、O形纹下会出现絮状斑（见图6-2-19）。

### （三）釉裂纹

#### 1. 釉裂纹概念

釉裂纹是元青花瓷器在烧制过程中和在后天环境影响下形成的，以直线、曲线等不规则线条形貌分布在釉表面的开裂纹路。

#### 2. 釉裂纹形成

釉裂纹的形成有以下几种原因：

（1）元青花瓷器在高温烧制时，釉表面的成分挥发造成空位，这些空位集聚成为空隙，因应力形成釉裂纹。

（2）釉层与微相之间膨胀不一可形成釉裂纹。

（3）由于$\equiv Si-O-R$中$R^+$和$H^+$的离子交换，$H^+$离子取代$R^+$位置，而$H^+$又比$R^+$小，因而产生应力，形成釉裂纹[102]。

（4）釉层与胎的膨胀系数不同，当温度变化时，胎釉之间会产生应力，由于胎较厚，承受应力的能力比釉层强，当应力大到一定程度时，釉层不能承受，便形成釉裂纹。

（5）釉层没有延展性，当外力打击超过其承受能力时，釉层便开裂形成釉裂纹。

釉裂纹未必与风化作用有直接的因果关系，有些釉裂纹是烧制过程中形成的，有的是后天环境影响下形成的。不是所有釉裂纹在显微成像下都能被观察到。在后天环境中，当温度发生变化或有可溶盐或难溶盐渗入而又析出时，产生的应力会使釉裂纹逐渐扩大。

陶瓷的釉表面出现裂纹，有时是人们利用胎与釉膨胀系数的差异，刻意追求的艺术效果。

元青花瓷器的釉裂现象不属于人们刻意追求的艺术效果。元青花瓷器上很少出现釉裂纹，如果有釉裂现象，绝大部分都属于工艺缺陷。

#### 3. 釉裂纹特征

釉裂纹有直线形貌，也有曲线形貌。对釉裂纹的具体描述如下：

粗细变化不明显（见图6-2-20、图6-2-21）。

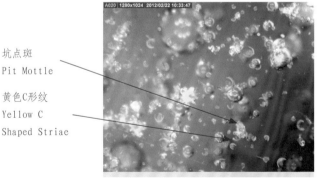

图6-2-17（Z1）

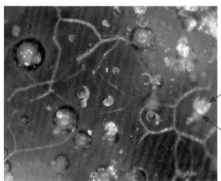

图6-2-18（T12）

图6-2-19（T26）

粗细有变化（见图6-2-22）。

粗细或颜色都没有明显变化（见图6-2-23）。

粗细或颜色都有明显变化（见图6-2-24、图6-2-25）。

一般来说，较早出现的釉裂纹偏粗，较晚出现的偏细，而且较早出现的变色情况比较晚出现的要严重。

在一条釉裂纹上会延伸出多条裂纹，像人参须一样（见图6-2-26）。

全部变色但无明显色阶变化（见图6-2-20），有的有色阶变化，颜色有白色、黄色、褐色、黑色等（见图6-2-21、图6-2-24、图6-2-25）。

部分变色（见图6-2-25）。

釉裂纹周边的特征包括：

色变泡或壁变泡的出现（见图6-2-27、图6-2-28）。

釉下变色（见图6-2-29、图6-2-30）。

釉下形成絮状斑（见图6-2-26、图6-2-31）。

这些特征可能单独或同时出现。

### （四）划痕

#### 1. 划痕概念

划痕是元青花瓷器釉表面与

粗细不明显，无明显色阶变化
Consistent Width, No Obvious Color Spectrum

图6-2-20（N5）

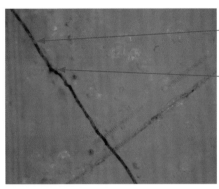
粗细不明显，有色阶变化
Consistent Width, Color Spectrum

图6-2-21（T36）

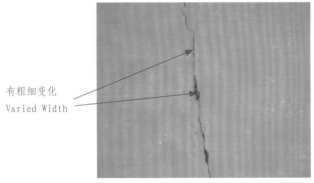
有粗细变化
Varied Width

图6-2-22（N3）

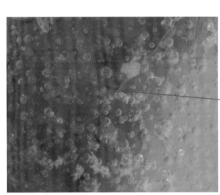
没有粗细、色阶变化
Consistent Width, No Color Spectrum

图6-2-23（T27）

较细釉裂纹
Thin Glaze Crack

较粗釉裂纹，有色阶变化
Thick Glaze Crack, Color Spectrum

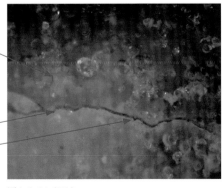
图6-2-24（T8）

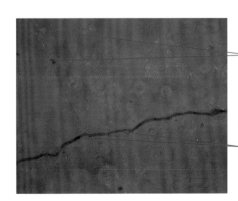
较细釉裂纹，部分变色
Thin Glaze Crack, Partial Color Spectrum

较粗釉裂纹，有色阶变化
Thick Glaze Crack, Color Spectrum

图6-2-25（T8）

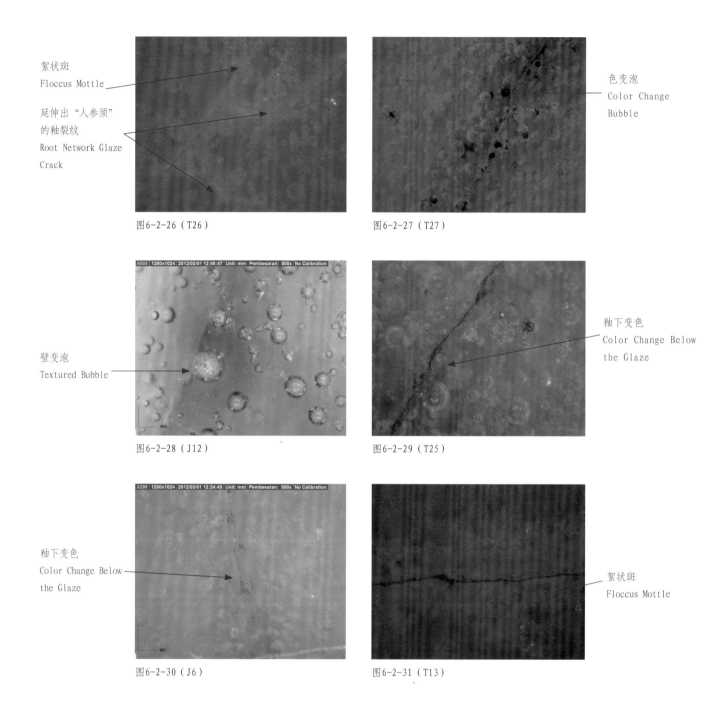

图6-2-26（T26） 图6-2-27（T27）
图6-2-28（J12） 图6-2-29（T25）
图6-2-30（J6） 图6-2-31（T13）

硬物碰划留下的破坏性纹路。

**2. 划痕形成**

元青花瓷器在后天长期使用中或存在环境中与硬物碰划，在釉表面会留下划痕。

**3. 划痕特征**

划痕的形貌特征：有粗有细，有长有短，有深有浅，有直线有曲线，颜色有白色、黄色、褐色等，随机无序地分布在釉表面（见图6-2-32—图6-2-42）。

划痕如牛毛，细而短，杂乱无章地分布（见图6-2-35、图6-2-36）。

在同一张图片上同时存在不同形貌（有的像毛毛虫）、不同颜色的划痕（见图6-2-35、图6-2-37、图6-2-38）。

划痕入釉较深，形成一定的凹槽，长期易积聚外来物质风化形成针眼斑（见图6-2-38、图6-2-39）。

划痕深入釉层，破坏了釉层结构，容易形成絮状斑（见图6-2-40—图6-2-42）。

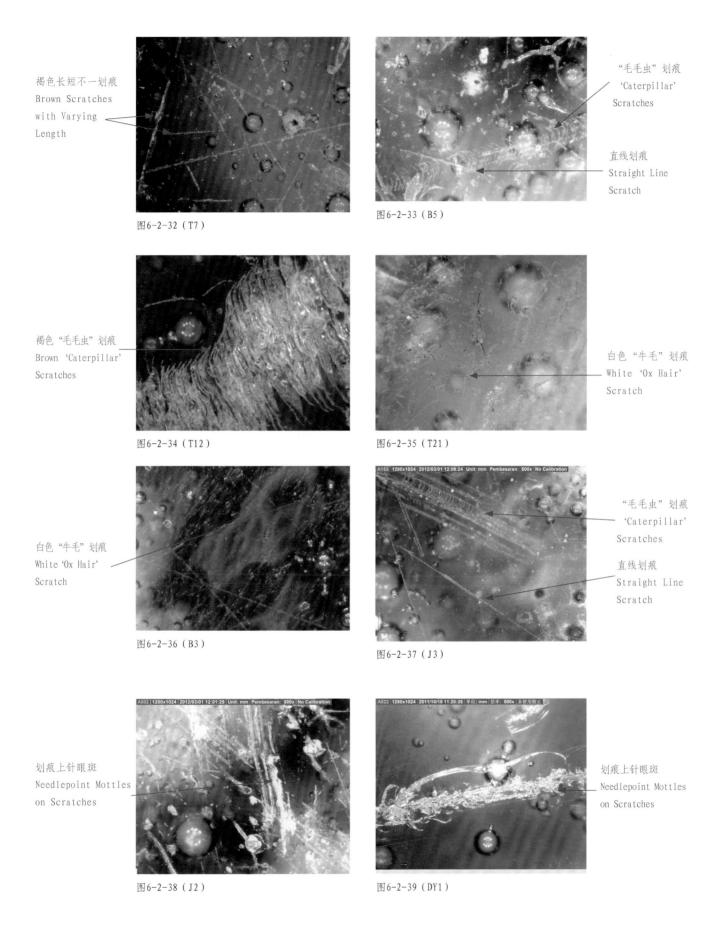

褐色长短不一划痕 Brown Scratches with Varying Length

图6-2-32（T7）

"毛毛虫"划痕 'Caterpillar' Scratches

直线划痕 Straight Line Scratch

图6-2-33（B5）

褐色"毛毛虫"划痕 Brown 'Caterpillar' Scratches

图6-2-34（T12）

白色"牛毛"划痕 White 'Ox Hair' Scratch

图6-2-35（T21）

白色"牛毛"划痕 White 'Ox Hair' Scratch

图6-2-36（B3）

"毛毛虫"划痕 'Caterpillar' Scratches

直线划痕 Straight Line Scratch

图6-2-37（J3）

划痕上针眼斑 Needlepoint Mottles on Scratches

图6-2-38（J2）

划痕上针眼斑 Needlepoint Mottles on Scratches

图6-2-39（DY1）

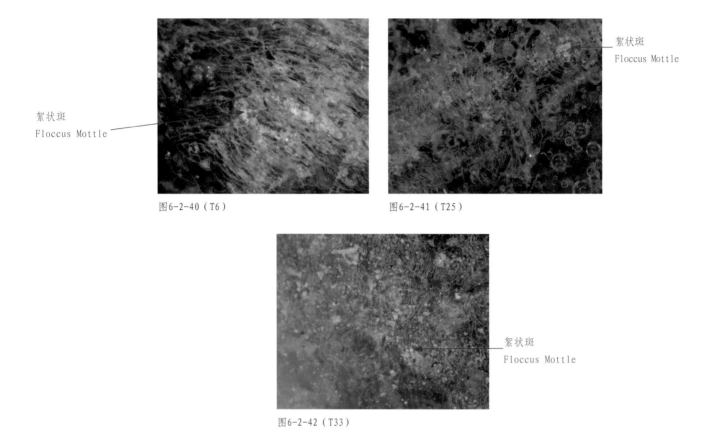

图6-2-40（T6）　　　　　图6-2-41（T25）

图6-2-42（T33）

# 第三节
## 元青花气泡类别

### 一、元青花气泡概念

元青花气泡是在元青花瓷器烧制过程中，在釉层中形成的气相迹型类别。

釉层中的气泡在烧制产生时，绝大多数都是透明无色的。

总体来说，元青花瓷器上气泡的大小及分布特征与明清时期青花瓷器上的气泡不一样。元青花上要么气泡很大，要么气泡较小，中等大小的气泡较少，而且元青花的气泡往往是三两个一组，零零星星，不会成群出现；明清时期青花瓷器上中等大小的气泡较多，分布密集（见图6-3-1—图6-3-6）。

### 二、元青花气泡种类

按气泡颜色变化与气泡内壁质地变化的特征，可将元青花气泡分为色变泡与壁变泡两大种类。

注：气泡只有内壁，没有外壁（见图6-3-7、图6-3-8）。

图6-3-1（N10）
元青花 YUB

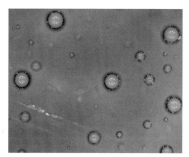

图6-3-2（N10）
元青花 YUB

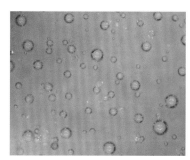

图6-3-3（K3）
明永乐青花明宣德青花 Ming Yongle

图6-3-4（K4）
明宣德青花 Ming Xuande

图6-3-5（DQ1）
清康熙青花 Qing Kangxi

图6-3-6（DQ2）
清雍正青花 Qing Yongzheng

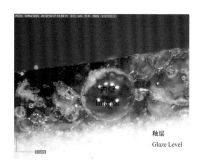

图6-3-7（DY1）

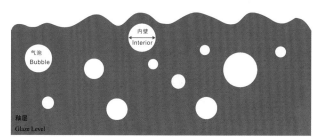

图6-3-8

# Section 3:
# The Yuan Dynasty Underglaze Blue Porcelain Air Bubble Category

## 1. YUB Air Bubble Defined

YUB Air Bubbles (also simplified as 'bubbles') are the trace category of gas phase appearances, formed during the firing process, within the glaze level of YUB porcelain.

Almost all bubbles formed during the firing process are clear and colorless.

The size, distribution, and properties of air bubbles within the glaze level of YUB are different from those found in later Ming and Qing Dynasty Underglaze Blue wares. The range in size from largest to smallest bubble in YUB is larger than that of later wares. Additionally, the distribution consists of primarily sparsely spread large and small bubbles, with medium size bubbles making up only a small minority. This is different from Ming and Qing Dynasty wares, which have a large portion of medium sized bubbles and a denser distribution in general (see pictures 6-3-1—6-3-6).

## 2. YUB Air Bubble Types

We use the color and internal texture changing properties of air bubbles to divide YUB air bubbles into two types: Color Change Bubble and Textured Bubble.

Note: Bubbles only have an interior wall, no exterior (see pictures 6-3-7 and 6-3-8).

### A. Color Change Bubble

#### I. Color Change Bubble Defined

Color Change Bubbles are the appearance of YUB porcelain air bubbles within the glaze level that have undergone a coloration change under the influences of the post-firing environment, but have retained their translucency.

#### II. Formation of Color Change Bubbles

For a bubble to change from clear transparency to having coloration, it must fulfill the following criterion:

a. Certain Amount of Iron

The change from a non-colored to colored bubble is primarily due to the reaction of Iron.

The vast majority of bubbles are clear directly after firing, and only a few of them – randomly distributed throughout the glaze level – contain trace amounts of elemental Iron. It is also possible, under the influence of the post-firing environment, for Iron to enter through channels into the glaze level and create colored bubbles.

b. Certain Amount of Water

Coloration is dependent on the presence of moisture. Not only can moisture help to carry foreign materials into the bubble, but also aides in the reactions that cause coloration.

c. Contact with the Outside Surroundings

Channels through the glaze level into the bubble interior can be:

i. Formed during the firing process:

When cooling from a molten state, the glaze is unable to fully recede over the mouth of an air bubble's escape passage, leaving a microscopic opening in the surface of the glaze. This opening acts as a channel between the bubble and the outside world.

The Firing process creates glaze crazes that act as channels.

## （一）色变泡

### 1. 色变泡概念

色变泡是元青花瓷器在后天环境影响下，釉层中透明气泡的颜色发生了变化，但仍具透影性的气泡（透影性即半透明状态）。

### 2. 色变泡形成

完整的无色透明气泡变成有色气泡需要以下条件：

（1）有一定含量的铁元素

气泡从无色变成有色，是气泡内有一定量的铁元素发生反应的结果。

刚烧制出来的元青花瓷器，釉层中的气泡绝大多数是无色透明的，在这些无色透明的气泡中，只有部分气泡内含有铁元素，这些气泡随机分布在釉层中。在后天环境的影响下，外界的铁元素也可经通道进入气泡内，使气泡变色。

（2）有一定的水分子

气泡从无色变成有色，离不开水的作用，水既可以是输送外界元素进入气泡的媒介，也可以参与反应使气泡变色。

（3）有连接外界环境的通道

釉层通往气泡内的通道可以有两种。

① 烧制过程中形成的

在烧制过程中形成的，有以下两种情况：

元青花高温烧制时，在熔融釉层冷却平复的过程中，釉未能将冲到釉表面气泡的破口完全填满，使釉表面留有微小孔隙，这些微小孔隙构成了气泡与外界环境连接的通道。

元青花烧制时形成的釉裂纹成为通道。

② 后天环境影响下形成的

釉层长期风化受到侵蚀，形成新的裂纹或孔隙，成为通道。

这些通道会随着时间的推移逐渐变大。

通道存在是气泡变色的关键因素，它可以使部分含铁元素的气泡变色，也可以让不含铁元素的气泡变色。

由于通道的存在，气泡与外界环境相通，外界环境中的各类物质可进入气泡内。

（4）有一定大的空间容量

气泡从无色变成有色需要一定大的空间容量。如果气泡内含一定量的铁元素，也有通道与外界环境相联，但气泡内却没有一定大的空间容量，气泡也不会变色。观察表明，极小的气泡基本是没有变色的。如果通道的空间足够大，通道也会变色，如釉裂纹的变色。

（5）经一定时间反应

气泡从无色变成有色的自然风化是一个较长的时间过程，因为通道极为细小。统计观察表明，部分存放在大气环境中的瓷器，至少40年左右的时间才有色变泡产生。40年之内的瓷器除了在烧制过程中产生的极个别灰泡之外，很少能观察到黄色、褐色等色变泡现象。

### 3. 色变泡种类

（1）非破口色变泡

① 非破口色变泡概念

非破口色变泡是用500X显微镜观察到的完整无破口的色变泡。

② 非破口色变泡种类

A. 局变泡

a. 局变泡概念

局变泡是整个气泡仅部分发生了颜色改变的非破口色变泡。

b. 局变泡特征

局变泡的颜色有黄色、褐色等（见图6-3-9—图6-3-14）。

部分局变泡在同一气泡内有色阶变化。

有的气泡仅仅是泡壁外延一圈改变了颜色，像土星光环围绕，气泡内颜色未变或部分变色。通常这类气泡都是泡壁外延一圈的颜色较深，越往气泡中心方向的颜色越淡（见图6-3-9、图6-3-10）。

有的气泡只有一处改变了颜色，变色部位随机（见图6-3-11）。

有的气泡部分改变了颜色，且与未变颜色的部分界限分明，没有渐变的形式，形成这种现象有两种原因：

一是气泡内壁出现裂纹，泡壁发生错位，形成了气泡内壁风化的屏障。这是一种特殊的局变泡（见图6-3-12）。

二是气泡内壁是凹凸不平的，凹凸不平的表面会形成气泡内壁风化的屏障（见图6-3-13）。

ii. Formed under the influences of the post-firing environment:

Over long periods of corrosive weathering, the glaze can present with new cracks and holes, these act as channels.

Some channels can gradually become larger over time.

The presence of these channels is necessary for the formation of color change bubbles; they allow for bubbles inherently containing Iron to change color, as well as allow external Iron to enter in and color the bubbles, post-firing.

Due to the presence of channels, some air bubbles are in contact with the external elements. This allows for the entrance of foreign materials into the bubble interior.

d. Certain Amount of Space

The discoloration of bubbles needs a certain amount of space. Even though a bubble may have trace amounts of Iron and channels to the external environment, if it does not have enough space, the coloration cannot occur. Observation has shown that minutely sized bubbles rarely change color. However, if the space within the channel is large enough, even the channel itself can become discolored: For example, the discoloration of glaze cracks.

e. Certain Amount of Time

The coloration of air bubbles through natural weathering is a relatively long process due to the minute size of the channels. Through several controlled observations, it has been found that pieces sitting in an open environment take approximately forty years to naturally develop color change bubbles. On pieces less than forty years of age, it is extremely rare to find yellow, brown, and so forth colored bubbles appearing (except for a trace amount of ashen colored bubbles sometimes created during the firing process).

III Color Change Bubble Types

a. Unbroken Mouth Color Change Bubbles

i. Unbroken Mouth Color Change Bubble Defined

Color change bubbles without an observable mouth under 500X magnification.

ii. Unbroken Mouth Color Change Bubble Types

(a) Semi-Changed Bubbles

(i) Semi-Changed Bubble Defined

An unbroken mouth color change bubble in which only a portion of the bubble has experienced discoloration.

(ii) Properties of Semi-Changed Bubbles

The colors of semi-changed bubbles can include: Yellow, brown, and other similar colorations (see pictures 6-3-9—6-3-14).

Some semi-changed bubbles can present with a color spectrum within a single bubble.

Some bubbles can present with full or partial coloration rings around the outside of the bubble, similar to the rings of Saturn; with the more external colors darker, and lightening towards the center (see pictures 6-3-9 and 6-3-10).

Some bubbles are partially discolored in a random manner (see picture 6-3-11).

Some bubbles have a clearly delineated boundary between the colored portion and the uncolored portion, the reasons for this being:

Cracks within the interior wall of the bubble act as a barrier for the further weathering of the bubble. This is a relatively rare type of semi-changed bubble (see picture 6-3-12).

The interior of the bubble is topographically uneven; these uneven areas act as barriers to the spread of weathering (see picture 6-3-13).

(b) Fully-Changed Bubbles

(i) Fully-Changed Bubble Defined

Fully-Changed Bubbles are unbroken mouth color change bubbles in which the entirety of the air bubble has changed color.

(ii) Properties of Fully-Changed Bubbles

Fully-changed bubbles primarily consist of yellow, brown, and black coloration and have some of the following characteristics:

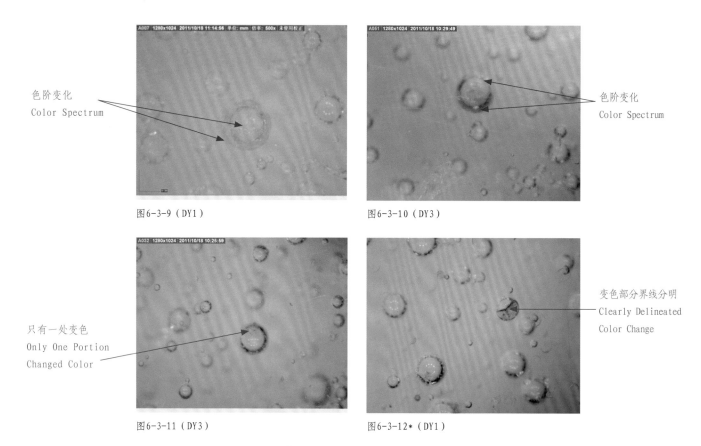

图6-3-9（DY1）　　　　　　　　　图6-3-10（DY3）

图6-3-11（DY3）　　　　　　　　　图6-3-12*（DY1）

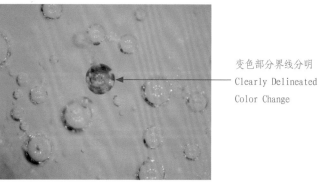

图6-3-13*（DY1）

B. 全变泡

a. 全变泡概念

全变泡是整个气泡颜色都发生了改变的非破口色变泡。

b. 全变泡特征

全变泡的颜色有黄色、褐色及黑色等。其主要特征为：

大部分全变泡在同一气泡内都有色阶变化（见图6-3-14、图6-3-15），部分气泡没有明显色阶变化（见图6-3-16）；

部分气泡内有龟裂纹路（见图6-3-17）；

部分距离较近的几个全变泡间有色差（见图6-3-18）；

部分距离较近的几个气泡都为全变泡，有时这些全变泡之间的玻璃相也会变色，形成连环全变泡（见图6-3-19、图6-3-20）。

C. 小结

在同一元青花瓷器上，局变泡与全变泡可以同时存在。

从无色泡变成色变泡、从局变泡变成全变泡是一个量变过程，这些气泡有色阶浓淡变化，泡与泡之间也会有色差变化。局变泡与全变泡可以同时存在，但它们均

---

*图片经过裁剪放大处理。

*Picture cropped and enlarged to show detail.

Majority of fully-changed bubbles present with a color variation (see pictures 6-3-14 and 6-1-15), while the remainder present with a single color (see picture 6-3-16).

Some of the bubble interiors have crazing (see picture 6-3-17).

Bubbles within close proximity can present with different colorations (see picture 6-3-18).

At times, when bubbles in close proximity change color, the glass phase between them also becomes discolored, creating a fully-changed bubble group (see pictures 6-3-19 and 6-3-20).

(c) Summary

A single piece of YUB porcelain can present with both semi and fully-changed colored bubbles.

Coloration is a process of quantitative change, a bubble can present with a wide spectrum of color, and a variety of colors can present within different bubbles. While semi and fully-changed bubbles can exist congruently, they must all retain translucency (see pictures 6-3-21 and 6-3-22).

The characteristics of bubble coloration shows that coloration happens as a result of the corrosive weathering of materials on the interior wall of each bubble – an indication of gradual weathering accumulation – and is not a filling of the entire bubble cavity.

In general, the area of a bubble with the darkest coloration is an indication of an area of glaze flaw; a channel offering contact with the external environment. This entry point is often the first to undergo weathering, and thus has the deepest coloration within the bubble.

Look at pictures 6-3-23 and 6-3-24 (where 6-3-23 is the two-dimensional and 6-3-24 is the three-dimensional view of the same spot on the same piece).

According to picture 6-3-23, we can see that area C of bubble B is darker than that of the rest of the bubble. From picture 6-3-24 we can see that area C is most likely the channel leading into bubble B. On bubble A of picture 6-3-23, we can only see a faint ring in picture 6-3-24, which indicates that the glaze surface is smooth and clean. This lets us know that bubble A probably lacks a channel directly to the glaze surface, and undergoes discoloration through the connection of area C on bubble B via passage D.

There are also color change bubbles with indistinct channels to the glaze surface.

Look at pictures 6-2-25 and 6-2-26 (where 6-3-25 is the two-dimensional and 6-3-26 is the three-dimensional view of the same spot on the same piece).

Bubble E in picture 6-2-25 is not visible but indicated by the red circle in picture 6-2-26. The glaze surface above bubble E is smooth and clean, indicating that the channel to the glaze surface is relatively small, and not observable under 500X magnification.

b. Broken Mouth Color Change Bubbles

i. Broken Mouth Color Change Bubble Defined

Broken mouth bubbles have an opening circumference, observable under 500X magnification, which does not exceed the largest circumference of the bubble (the observable mouth is smaller than the widest part of the body).

Currently on YUB samples, we have yet to discover Broken Mouth Color Change Bubbles (all are textured bubbles).

## B. Textured Bubbles

I. Textured Bubble Defined

Textured Bubbles are the appearance of YUB porcelain bubbles in which the interior has undergone weathering under the influence of the post-firing environment; gaining texture and losing their inherent translucency.

To differentiate whether or not a bubble has undergone texturing, we must look to see if that bubble has retained its translucency. If the bubble has undergone weathering that causes a loss of translucency, it belongs to the textured bubble type.

II. Textured Bubble Types

a. Unbroken Mouth Textured Bubbles

i. Unbroken Mouth Textured Bubble Defined

These are textured bubbles that do not have an opening

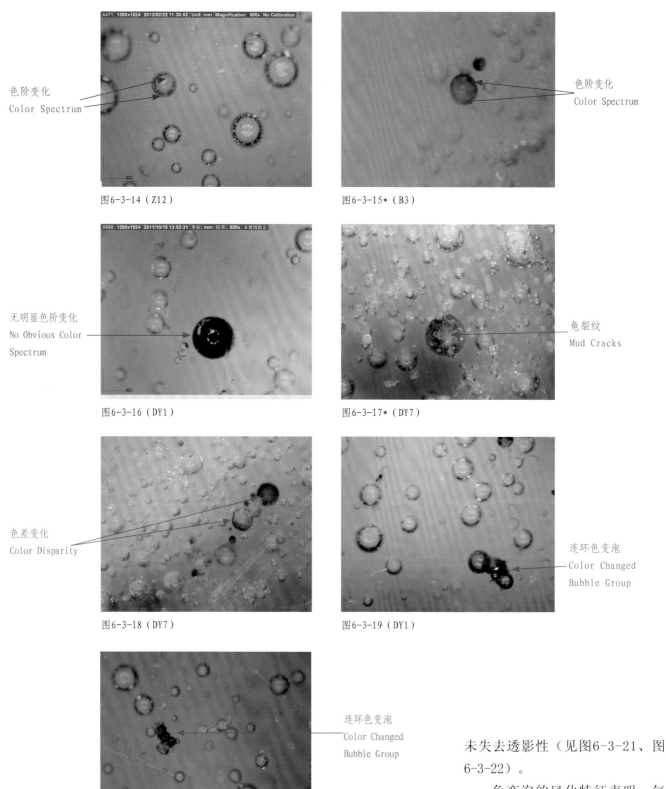

图6-3-14（Z12） 图6-3-15*（B3）

图6-3-16（DY1） 图6-3-17*（DY7）

图6-3-18（DY7） 图6-3-19（DY1）

图6-3-20（T3）

未失去透影性（见图6-3-21、图6-3-22）。

色变泡的风化特征表明，气泡内的风化过程是侵蚀物紧贴气泡的内壁，以积累的模式渐进风化，而不是弥漫在气泡内的空间随机风化。

通常色变泡哪个部位颜色深，说明那里的釉层结构有缺

---

*图片经过裁剪放大处理。
*Picture cropped and enlarged to show detail.

observable under 500X magnification.

ii. Unbroken Mouth Textured Bubble Types

Unbroken mouth textured bubbles assume white, yellow, brown, and black coloration, with some bubble interiors undergoing full texture change and some only experiencing partial change. According to the interior topography and variation in weathering intensity, unbroken mouth textured bubbles can assume several different types of appearance. In accordance with the variances in appearance, we can divide them into the following three types:

(a) Foggy Needlepoint Textured Bubbles

Foggy needlepoint textured bubbles are ones in which the interior texture has weathered into an even foggy needlepoint texture.

The distribution of needlepoints is even and a high degree of weathering causes a dense covering, in general the glass phase is no longer visible. Some foggy needlepoint textured bubbles present with a color spectrum, some do not; with the majority belonging to the fully changed unbroken mouth textured bubble variety (see pictures 6-3-27 and 6-3-28).

(b) Frosted Needlepoint Texture Bubbles

Frosted needlepoint textured bubbles are ones in which the interior texture has weathered into a frosted texture, and in which the pits are larger than that of the foggy needlepoint texture. Frosted needlepoint textured bubbles represent a higher severity of weathering in which, most of the time, the glass phase is no longer visible, and the bubble interior retains a relatively even texture. Some of these bubbles have a color spectrum, some do not; with the majority also belonging to the fully changed unbroken mouth textured bubble variety (see pictures 6-3-29 — 6-3-32).

(c) Grainy Needlepoint Textured Bubbles

Grainy needlepoint textured bubbles are ones in which the interior texture has weathered into a rough pitted texture. For the most part, needlepoints retain independence and the glass phase is visible between them. The majority of these bubbles have some color spectrum, and belong to the fully changed unbroken mouth textured bubble variety (see pictures 6-3-33 — 6-3-36).

(d) Formation of Foggy Needlepoint, Frosted Needlepoint and Grainy Needlepoint Texture Bubbles

The glaze surface of ceramics has a random distribution of cracks and openings. Additionally, soil has a certain amount of acidity or alkalinity that acts as a corrosive on ceramic glaze creating more glaze level cracks and openings. Materials from the external environment can traverse these cracks and openings to enter into the air bubbles and deposit soluble and partially soluble salts. In environments of alternating moisture, partially soluble salts are not easily dispensed from the air bubbles. Soluble salts and small dust particles crystallize during the evaporation of water and become deposited on the bubble interior, thus creating textured bubbles.

The positive and negative ions within soil and within the glaze level can partake in reactions within the bubble interior and leave soluble deposits of varying colors, thus the differing colors of textured bubbles.

Similar to color change bubbles, there is a pattern of the darker areas of a bubble indicating an area of structural flaw and contact with the external environment. Therefore, the points of channels leading into textured bubbles are the areas of first change and deepest coloration.

Look at pictures 6-3-37 and 6-3-38 (where 6-3-37 is the two-dimensional and 6-3-38 is the three-dimensional view of the same spot on the same piece).

Bubble F in picture 6-3-37 is hardly visible in picture 6-3-38 and marked with a red circle. Additionally, the point indicated by arrow G is clearly more visible in the picture 6-3-38, which is an indication of the presence of a broken mouth on the surface of the glaze. The surrounding glaze is relatively smooth and clean.

From this we can tell: Weathering has changed the interior of bubble F. The broken mouth indicated by arrow G is a point of glaze weakness, and is the channel between the external environment and the bubble within the glaze level. Foreign materials are able to enter bubble F through the channel, causing it to lose its transparency thus forming a textured bubble.

陷、有连接外界的通道存在。所以，通道在色变泡内的出口处，一般就是颜色最早变化和颜色较深的地方。

见图6-3-23、图6-3-24（图6-3-23、图6-3-24为同一器物同一位置图片，图6-3-23为二维图片，图6-3-24为三维图片）。

如图6-3-23中，气泡B上C处的颜色比该气泡其他部位的颜色要深，从图6-3-24可看出，通往气泡B内部的通道很有可能位于C处，而图6-3-23中的气泡A，在图6-3-24中却隐约只见淡淡的轮廓，气泡A所在处的釉表面光滑洁净，这说明气泡A之所以变色，极有可能是气泡B上C处通道连接D处通道造成的。

也有的色变泡，在釉表面的通道变色不明显。

见图6-3-25、图6-3-26（图6-3-25、图6-3-26为同一器物同一位置图片，图6-3-25为二维图片，图6-3-26为三维图片）。

图6-3-25中的气泡E，在图6-3-26中位于红圈釉下处，不可见。气泡E所在处的釉表面光滑洁净，说明釉表面通往气泡E的通道

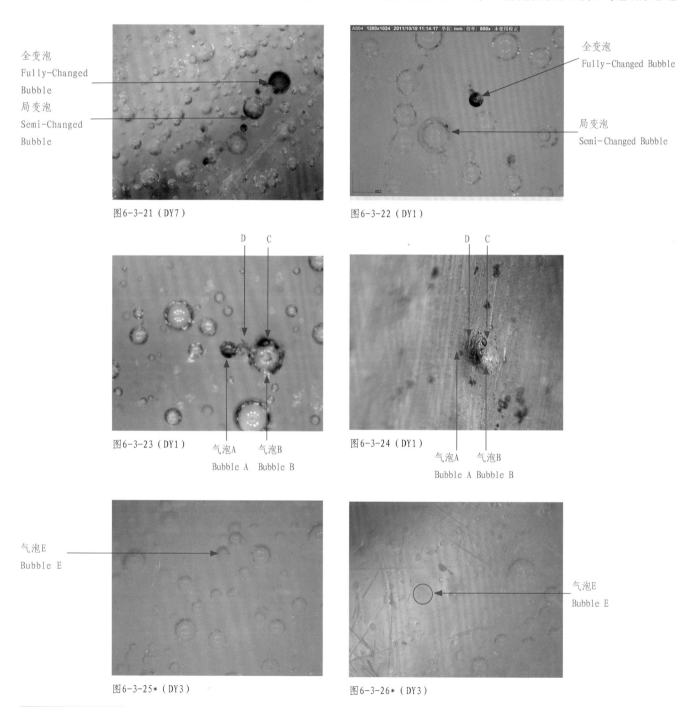

图6-3-21（DY7）　　　　　图6-3-22（DY1）

图6-3-23（DY1）　　　　　图6-3-24（DY1）

图6-3-25*（DY3）　　　　　图6-3-26*（DY3）

*图片经过裁剪放大处理。
*Picture cropped and enlarged to show detail.

Look at pictures 6-3-39 and 6-3-40 (where 6-3-39 is the two-dimensional and 6-3-40 is the three-dimensional view of the same spot on the same piece).

Bubble H in picture 6-3-39 is a faintly visible ring in picture 6-3-40, indicating that it is the interior of the bubble, not the glaze surface, that has undergone mottling. The glaze surrounding bubble H is relatively clean, but there is a presence of scrapes across the surface; these are areas of glaze weakness. Additionally, arrows I and J point to areas on the scrapes of darker coloration, and most likely represent the channels between the external environment and the bubble interior. Foreign materials are able to pass through these channels into bubble H causing it to lose its transparency thus forming a textured bubble.

Look at pictures 6-3-41 and 6-3-42 (where 6-3-41 is the two-dimensional and 6-3-42 is the three-dimensional view of the same spot on the same piece).

Bubble K in picture 6-3-41 is a faintly visible ring in picture 6-3-42, indicating that it is the interior of the bubble that has undergone mottling. The glaze surrounding bubble K is relatively clean, but there is clearly a presence of scrapes across the surface; these are areas of glaze flaw as indicated by arrow L. They most likely created the channel between the external environment and the bubble interior, allowing foreign materials to pass into bubble K, causing it to lose its transparency, thus forming a textured bubble.

(e) Webbed Bubbles

Textured bubbles in which the interior texture has weathered into spider web shape appearance. The bubble interior has spider web shaped striae and yellow coloration without color variation (see picture 6-3-43).

Webbed bubbles are very rare. According to picture 6-3-43, we can see that this bubble is located on a glaze crack, which suggests the crack is the medium for contact with the external environment. According to the interior topographical unevenness and varying degrees of weathering, certain deposits have formed spider web shaped striae on the interior of the bubble. Also, due to the proximity of the bubble to the blue dye, the amount of Iron is relatively high; once oxidized,

yellow coloration has occurred.

b. Broken Mouth Textured Bubbles

i. Broken Mouth Textured Bubble Defined

Broken mouth bubbles have an opening circumference, observable under 500X magnification, that does not exceed the largest circumference of the bubble (the observable mouth is smaller than the widest point of the bubble).

Note: Conceptually, some bubble mouths are microscopically small, unobservable under 500X magnification. These are unobservable-broken-mouth 'broken mouth bubbles'. For the time being, these are classified as the unbroken mouth bubble type.

ii. Formation of Broken Mouth Textured Bubbles

The formation of broken mouth textured bubbles is similar to that of color change and unbroken mouth textured bubbles (Refer to Chapter 6 Section 3 Formation of Color Change Bubbles and Formation of Unbroken Mouth Textured Bubbles).

The weathering occurring in the bubble interior corrodes the internal glaze level; additionally, the weathering occurring on the glaze surface can also work to corrode towards the bubble. Thus, the formation of broken mouth bubbles can be divided into two types:

(a) Single Direction

During the firing of YUB porcelain as molten glaze cools, if air bubbles come near to the surface of the glaze without breaking and the temperature cools relatively quickly, the air within the bubble contracts. This causes a small crater on the surface of the glaze, possibly lower than the surrounding glaze surface (see picture 6-3-44), or higher than the surrounding glaze surface (see picture 6-3-45). This indention tends to collect sediments, which corrodes the glaze.

During the firing of YUB porcelain as molten glaze cools, if air bubbles burst through the surface of the glaze and the temperature drops relatively quickly or the glaze is relatively viscous, the mouth of the bubble is unable to close, resulting in a tiny hole in the surface of the glaze. However,

较小，用500X显微镜难以观察到。

（2）破口色变泡

① 破口色变泡概念

破口色变泡是用500x显微镜能观察到的，其破口直径小于气泡直径的色变泡。

目前在元青花标本上尚未发现破口色变泡（破口泡均为壁变泡）。

（二）壁变泡

1. 壁变泡概念

壁变泡是元青花瓷器在后天环境影响下，釉层中的气泡内壁质地发生了风化现象，被风化处失去了透影性的气泡。

区分气泡内壁是否发生质变的依据就是气泡是否失去了透影性。如果气泡有部分风化处不具有透影性，就属于壁变泡。

2. 壁变泡种类

（1）非破口壁变泡

① 非破口壁变泡概念

非破口壁变泡是用500x显微镜能观察到的完整无破口的壁变泡。

② 非破口壁变泡分类

非破口壁变泡颜色有白色、黄色、褐色及黑色等，有的为全变，即整个内壁质地发生变化，有的为局变，仅局部发生变化。由于气泡内壁的凹凸不平，风化作用多样，非破口壁变泡质地也呈现多种不同的形貌。根据非破口壁变泡内部的风化形貌，可将其分成以下三种：

A. 雾状针眼型

雾状针眼型非破口壁变泡，内壁的风化以雾状针眼斑形式呈现。针眼与针眼之间分布均匀，通常看不见玻璃相。有的没有明显色阶变化，有的有色阶变化。大部分是全变非破口壁变泡（见图6-3-27、图6-3-28）。

B. 磨砂针眼型

磨砂针眼型非破口壁变泡，内壁的风化以磨砂针眼状形式呈现，其针眼要比雾状型粗砺，是风化强度较雾状针眼型更高的一类。通常也看不见玻璃相，泡内质地显得较均匀。有的没有明显色阶变化，有

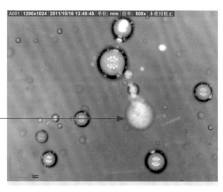

雾状针眼型
无明显色阶变化
Foggy Needlepoint Bubble, No Obvious Color Spectrum

图6-3-27（DY1）

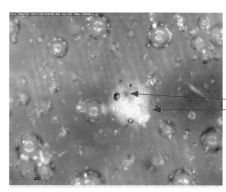

雾状针眼型
有色阶变化
Foggy Needlepoint Bubble with Color Spectrum

图6-3-28（DY7）

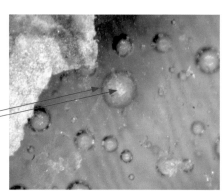

磨砂针眼型
有色阶变化
Frosted Needlepoint Bubble with Color Spectrum

图6-3-29（T2）

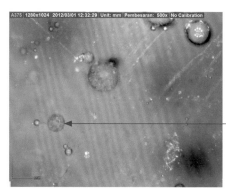

磨砂针眼型
无明显色阶变化
Frosted Needlepoint Bubble, No Obvious Color Spectrum

图6-3-30（J8）

these openings are microscopic and often unobservable under 500X magnification. Over a long period of weathering, the glaze surface corrodes, and the mouth of the bubble progressively enlarges (see picture 6-3-46).

Under the above stated two types of glaze surface corrosion, the final result is a broken mouth bubble. Materials from the external environment can easily enter and weather the bubble, forming a broken mouth textured bubble.

(b) Double Direction

Similar to single direction, while one of the two types of glaze surface level corrosion is occurring; simultaneously, because of alternate existing channels, internal weathering also occurs. Internal and external double directional corrosion thins the bubble wall and eventually creates a bubble mouth. At the same time, the interior of the bubble is undergoing textural changes; the result is a broken mouth textured bubble.

iii. Broken Mouth Textured Bubble Types

(a) Round Mouth Textured Bubbles

(i) Round Mouth Textured Bubble Defined

A broken mouth textured bubble with a circular mouth opening.

(ii) Properties of Round Mouth Textured Bubbles

The majority of broken mouth bubble openings, regardless of opening size, have a smooth and even edge.

The interior of round mouth bubbles is similar to that of other textured bubbles (refer to Chapter 6 Section 3 Unbroken Mouth Textured Bubble Types). They present with frosted needlepoint texture (see pictures 6-3-47—6-3-50), as well as frosted and grainy needlepoint texture (see picture 6-3-51).

Some round mouth bubble interiors can become fully-changed (see pictures 6-3-47—6-3-49); while others only become partially-changed, with a certain portion of the glass phase still visible (see pictures 6-3-50 and 6-3-51).

It is common for bits to become lodged inside the round mouth bubble causing visual blockage of the glass phase below (see pictures 6-3-48 and 6-3-49). Some bubbles can present with a color spectrum (see pictures 6-3-48—6-3-50); while others do not (see picture 6-3-47).

(b) Irregular Mouth Textured Bubbles

(i) Irregular Mouth Textured Bubble Defined

A broken mouth textured bubble with an irregularly shaped mouth opening.

(ii) Properties of Irregular Mouth Textured Bubbles

Even though the shape of the mouth opening is irregular, the edges retain a smooth evenness, without 'teeth' around the edges. Bubbles presenting with jagged, teeth-like edges are usually newly formed broken mouth bubbles.

The interior of an irregular mouth bubble is similar to that of other textured bubbles (refer to Chapter 6 Section 3 Unbroken Mouth Textured Bubble Types). They present with frosted needlepoint texture (see pictures 6-3-52 and 6-3-53), as well as grainy needlepoint texture (see pictures 6-3-54—6-3-57).

Some irregular mouth bubble interiors can become fully-changed (see pictures 6-3-52 and 6-3-53); while others only become partially-changed, with a certain portion of the glass phase still visible (see pictures 6-3-54, 6-3-56 and 6-3-57).

It is common for bits to become lodged inside the irregular mouth bubble causing visual blockage of the glass phase below (see pictures 6-3-52 and 6-3-54). Some bubbles can present with a color spectrum (see pictures 6-3-53, 6-3-55, and 6-3-57); while others do not (see pictures 6-3-52 and 6-3-56).

(c) Summary

On a single piece of YUB porcelain, round and irregular mouth bubbles can present simultaneously. The majority of the mouth edges present with a certain smoothness; with a higher percentage of smooth openings being a mark of quantitative trace change.

According to the study samples, portions of the broken mouth bubbles have color variation. Regardless of whether the bubble interior is partially or fully covered by foreign materials, coloration begins at the bubble mouth; top darker, bottom lighter. This is due to the fact that the open mouth is the channel to the external environment, and coloration begins

的有色阶变化。少部分是局变，大部分属于全变非破口壁变泡（见图6-3-29—图6-3-32）。

C. 颗粒针眼型

颗粒针眼型非破口壁变泡，内壁的风化以颗粒针眼斑呈现。针眼与针眼之间较独立，还能看见玻璃相。它们大多有色阶变化，少部分是局变，大部分属于全变非破口壁变泡（见图6-3-33—图6-3-36）。

D. 雾状针眼型、磨砂针眼型、颗粒针眼型非破口壁变泡形成

陶瓷釉表面随机分布有裂纹或孔隙。此外，土壤中具有一定酸或碱性的水溶液会侵蚀元青花瓷器釉表面，使釉表面形成裂纹或孔隙。外界物质通过这些裂纹和孔隙进入气泡内，在气泡内生成可溶盐及难溶盐。在干湿交替的环境中，气泡中的难溶盐难以流失，可溶盐及一些细小灰尘，在水蒸发后的气泡中形成结晶，它们在气泡内壁叠加沉积，形成壁变泡。

土壤中的及釉层中的阴阳离子会在气泡内发生反应，形成不同颜色的难溶盐沉积物，所以壁变泡有多种颜色。

与色变泡颜色变化规律类似，通常壁变泡哪个部位颜色深，说明那里的釉层结构有缺陷，有连接外界的通道存在。所以，通道在壁变泡内的出口处，一般就是颜色最早

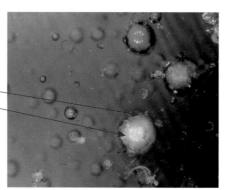
磨砂针眼型
有色阶变化
Frosted Needlepoint Bubble with Color Spectrum

磨砂针眼型
无明显色阶变化
Frosted Needlepoint Bubble, No Obvious Color Spectrum

图6-3-31（T35）　　　　　图6-3-32（J3）

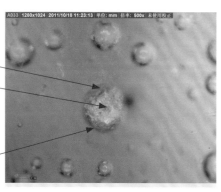
颗粒针眼型
有色阶变化
Grainy Needlepoint Bubble with Color Spectrum Bubble
可见玻璃相
Glass Phase Visible

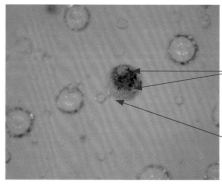
颗粒针眼型
有色阶变化
Grainy Needlepoint Bubble with Color Spectrum
可见玻璃相
Glass Phase Visible

图6-3-33（DY1）　　　　　图6-3-34（DY1）

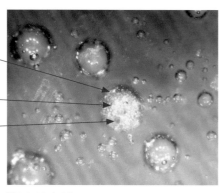
可见玻璃相
Glass Phase Visible
颗粒针眼型有色阶变化
Grainy Needlepoint Bubble with Color Spectrum

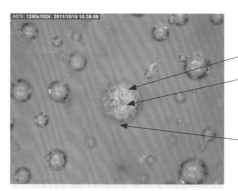
颗粒针眼型
有色阶变化
Grainy Needlepoint Bubble with Color Spectrum
可见玻璃相
Glass Phase Visible

图6-3-35（N2）　　　　　图6-3-36（DY3）

at this point (see pictures 6-3-58 and 6-3-59). Some bubbles assume an indistinct color variation.

Look at pictures 6-3-60 and 6-3-61 (where 6-3-60 is the two-dimensional and 6-3-61 is the three-dimensional view of the same spot on the same piece).

Bubble M in picture 6-3-60 is hardly visible in picture 6-3-61 and marked with a red circle. Additionally, the point indicated by arrow N is clearly more visible in picture 6-3-61, which is an indication of the presence of a broken mouth on the surface of the glaze. The surrounding glaze is relatively smooth and clean.

From this we can tell: Weathering has changed the interior of bubble M. The broken mouth indicated by arrow N is a point of glaze weakness, and is most likely a channel between the external environment and the bubble within the glaze level. Foreign materials are able to enter bubble M through the channel, causing it to lose its transparency thus forming a textured bubble.

This all goes to show, bubble corrosion is a gradual weathering process that affects the internal bubble surface; and it is not a filling of the bubble cavity through sudden weathering. The area around the bubble mouth presents with the darkest coloration; because it is the channel that connects to the external environment. Therefore, the area around the bubble mouth is the first to undergo coloration, and receives comparatively deeper coloration.

### c. Summary

On a single piece of YUB porcelain, it is possible for there to be a congruent appearance of unbroken mouth and broken mouth textured bubbles. The weathering severity on the inside of unbroken mouth textured bubbles is usually less than that of broken mouth bubbles, which is consistent with the passage of time principle.

## C. Faux Textured Bubbles

### I. Faux Textured Bubbles Defined

Faux Textured Bubbles are the appearance of needlepoint, pit, or overlay mottles, caused by natural weathering or man-made aging, presenting directly over an air bubble within the glaze level; they are easily misconstrued as a textured bubble.

### II. Distinguishing Faux Textured Bubbles

There are three methods for distinguishing faux and real textured bubbles:

Diameter Comparison Method;

Focus Selection Method;

Bunney Imaging Method.

When the diameter of the mottle is larger than the diameter of the air bubble we can use the Diameter Comparison or Bunney Imaging Method to differentiate; when the diameter of the mottle is smaller than that of the air bubble we can use the Focus Selection or Bunney Imaging Method to differentiate.

### a. Diameter of the Mottle Larger than the Bubble

Using the Diameter Comparison Method: When the diameter of the air bubble is smaller than that of the mottle, and the mottle completely covers the air bubble; we can directly distinguish this bubble as the faux textured bubble type (see picture 6-3-62 bubble O).

If the diameter of the mottle is larger than the air bubble, but does not completely cover the bubble below; we must utilize the Bunney Imaging Method to further distinguish the true type of the bubble.

Look at pictures 6-3-63 and 6-3-64 (where 6-3-63 is the two-dimensional and 6-3-64 is the three-dimensional view of the same spot on the same piece).

The textured bubble P in picture 6-3-63 is hardly visible in picture 6-3-64, and marked with a red circle. Whereas the dark colored mottle indicated by arrow Q in picture 6-3-63 is also visible in picture 6-3-64, with shape and size the same as picture 6-3-63.

The bubble P in picture 6-3-63 is a valid unbroken mouth textured bubble. This is because the texture is on the bubble interior and not on the glaze surface. The dark colored mottle indicated by arrow Q in both pictures is an identical point on the pictured bubble. This indicates that mottle indicated at point Q is on the glaze surface, and most likely a glaze flaw and a channel connecting the external environment with the

变化和颜色较深的地方。

见图6-3-37、图6-3-38（图6-3-37、图6-3-38为同一器物同一位置图片，图6-3-37为二维图片，图6-3-38为三维图片）。

图6-3-37中的气泡F，在图6-3-38中位于红圈釉下处，不可见。图6-3-37中箭头G所指深色处，在图6-3-38中更清晰可见，实为釉表面的破口，破口周围釉表面较干净，且无明显变色。

由此可推断，气泡F内壁有风化形成的斑块。箭头G所指的破口，属于釉表面的结构缺陷，是外界连接气泡内的通道。外界物质通过釉表面的通道进入釉层中的气泡F内，使气泡内壁发生质地变化，失去了透影性，形成壁变泡。

见图6-3-39、图6-3-40（图6-3-39、图6-3-40为同一器物同一位置图片，图6-3-39为二维图片，图6-3-40为三维图片）。

图6-3-39中的气泡H，在图6-3-40中却隐约只见淡淡的轮廓，说明风化斑块主要存在于气泡H内壁。气泡H所在处的釉表面比较洁净，但有较多划痕，划痕属于釉表面的结构缺陷。划痕上有几处颜色较深，如箭头I、箭头J所指处。它们很有可能是是外界连接气泡内的通道，外界物质通过釉表面的通道进入釉层中的气泡H内，使气泡内壁发生质地变化，失去了透影性，形成壁变泡。

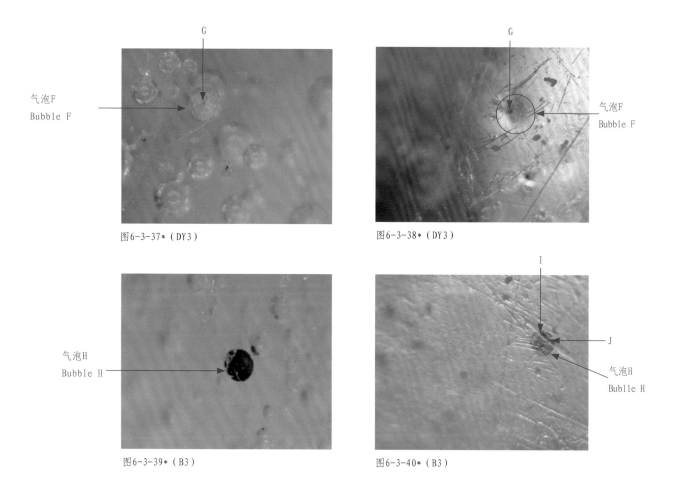

图6-3-37*（DY3）　　　　图6-3-38*（DY3）

图6-3-39*（B3）　　　　图6-3-40*（B3）

*图片经过裁剪放大处理。
*Picture cropped and enlarged to show detail.

bubble interior.

In picture 6-3-64, aside from the dark mottle indicated by arrow Q, the surrounding glaze level is essentially smooth. This shows that the textured bubble has undergone a slow, long, quantitative change process. Foreign materials have continuously passed through the channel in the glaze level and entered into the glaze level air bubble, causing a texture change and loss of translucency.

b. Diameter of the Mottle Smaller than the Bubble

When the mottle on the air bubble has a diameter that is smaller than that of the air bubble, we can utilize the Focus Selection Method to differentiate.

When an appropriate focus has been selected for clearly observing said mottling, we must notice what point of the observed object is most clear, any portions of the object that are on a different depth will gradually become blurred. If we observe one or more mottle points with clarity at different depths it tells us one of two things:

First: Part of the mottle is in the air bubble interior, and another part is on the surface level;

Second: Both points of the mottle are within the air bubble, but at different depths.

The Focus Selection Method is usually only suitable for air bubbles with a larger size. If the bubble is too small, it is difficult to differentiate clarity at different depths of the bubble. In this case, we can use the Bunney Imaging Method to help differentiate.

Due to the small size of the air bubble in question pictured in 6-3-65, it is too difficult to utilize the Focus Comparison Method, thus the Bunney Imaging Method must be used.

Look at pictures 6-3-65 and 6-3-66 (where 6-3-65 is the two-dimensional and 6-3-66 is the three-dimensional view of the same spot on the same piece).

Mottle R in picture 6-3-65, is clearly visible on the glaze surface in picture 6-3-66. This is to say that the 'texture' of mottle R is actually only present on the glaze surface and not the bubble interior. Therefore this bubble is classified as a faux textured bubble.

Look at pictures 6-3-67 and 6-3-68 (where 6-3-67 is the two-dimensional and 6-3-68 is the three-dimensional view of the same spot on the same piece).

The textured bubble S in picture 6-3-67 is not visible in picture 6-3-68, but is actually present below glaze at the position marked with a red circle. The dark colored mottle indicated by arrow U on picture 6-3-67, becomes invisible on picture 6-3-68. Meanwhile the dark colored mottle indicated by arrow T, is however visible on picture 6-3-68.

Mottle U is not on the glaze surface, but rather inside bubble S; formed by weathering on the bubble interior. Mottle T is present on the surface of the glaze, and is most likely a glaze flaw that was formed by weathering directly on the glaze face. It is a channel between the external environment and the bubble interior, allowing for the passage of foreign materials into the bubble, and creating a textural change to the bubble interior; losing its translucency and becoming a textured bubble.

见图6-3-41、图6-3-42（图6-3-41、图6-3-42为同一器物同一位置图片，图6-3-41为二维图片，图6-3-42为三维图片）。

图6-3-41中的气泡K，在图6-3-42中却隐约只见淡淡的轮廓，说明风化斑块主要存在于气泡K内壁。气泡K所在处的釉表面比较洁净，但有一处较明显的划痕，如箭头L所指。划痕属于釉表面的结构缺陷，它很有可能是外界连接气泡内的通道，外界物质通过釉表面的通道进入釉层中的气泡K内，使气泡内壁发生质地变化，失去了透影性，形成壁变泡。

E. 蛛网壁变型

蛛网壁变型非破口壁变泡，内壁的风化纹路以蜘蛛网纹形式呈现。颜色为黄色，无色阶变化（见图6-3-43）。

蛛网壁变泡较罕见。由图6-3-43可见，气泡所处上方的釉表面存在釉裂纹，由此可推断釉裂纹为连接外界与气泡内部的通道之一。由于气泡内壁凹凸不平，风化作用紧贴气泡内壁进行，气泡内壁凹槽处风化物质堆积，从而形成如蜘蛛网般的纹路。由于气泡位于青料处，青料含铁氧化物，铁氧化物的风化使气泡颜色变为黄色。

（2）破口壁变泡

① 破口壁变泡概念

破口壁变泡是用500x显微镜能观察到的，其破口直径小于气泡直径的壁变泡。

注：从理论上分析，有的气泡破口极其微小，即使使用500x显微镜也观察不到。我们将这些观察不到破口的"破口泡"，暂时都列入非破口泡之类。

② 破口壁变泡形成

破口壁变泡的形成原因部分与色变泡以及非破口壁变泡的形成原因相似（参见本章本节色变泡形成及非破口壁变泡形成）。

气泡内壁的风化会对气泡内壁的釉层造成侵蚀，而气泡所在处的釉表面的风化也会对釉层造成侵蚀，所以我们可将破口壁变泡的形成分成两类：

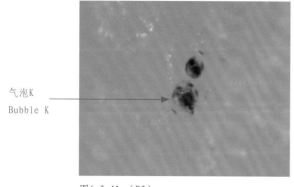

图6-3-41*（B3）

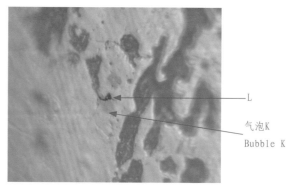

图6-3-42*（B3）

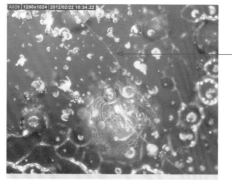

图6-3-43（Z1）

\*图片经过裁剪放大处理。
\*Picture cropped and enlarged to show detail.

A．单向侵蚀

元青花高温烧制时，在熔融釉层冷却平复的过程中，当气泡升至釉表面而又未冲破釉表面时，如果降温较快，气体体积缩小，气泡上方的釉表面会形成凹槽，或低于釉表面（见图6-3-44）或高出釉表面（见图6-3-45）。凹槽易长期积垢从而侵蚀釉表面。

元青花高温烧制时，在熔融釉层冷却平复的过程中，有些升至釉表面的气泡，由于降温较快或釉层粘度较大而没有闭合，从而使釉表面出现微小的孔洞。不过这些孔洞过于微小，用500倍显微镜一般是观察不到的。长期的风化作用侵蚀釉表面，使孔洞逐渐变大（见图6-3-46）。

釉表面这两种形式的风化作用不断侵蚀釉表面，最终导致气泡上方的釉表面出现较大破口，形成破口泡，外界物质更易进入产生风化作用，形成破口壁变泡。

B．双向侵蚀

双向侵蚀，一方面同单向侵蚀的情况类似，另一方面，气泡内壁因通道存在也会产生风化作用。内外双向的风化作用会使气泡壁变薄，从而产生破口；同时，气泡内壁的风化也使气泡内壁的质地发生改变，最终形成破口壁变泡。

③ **破口壁变泡种类**

A．圆口破口壁变泡

a．圆口破口壁变泡概念

圆口破口壁变泡是破口为圆形的壁变泡。

b．圆口破口壁变泡特征

大部分气泡的破口，无论大小，其破口边沿都是平滑的。

圆口破口壁变泡内壁的风化形貌与非破口壁变泡的风化形貌类似（参见本章本节非破口壁变泡种类），有磨砂针眼型（见图6-3-47—图6-3-50），也有磨砂针眼与颗粒针眼并存的壁变形式（见图6-3-51）。

有的圆口破口壁变泡为全变（见图6-3-47—图6-3-49）。有的为局变，透过气泡还能见玻璃相（见图6-3-50、图6-3-51）。

有的气泡破口下方可见物质存在，不可见玻璃相（见图6-3-48、图6-3-49）；有的气泡内壁有色阶变化（见图6-3-48、图6-3-50），有的没有明显色阶变化（见图6-3-47）。

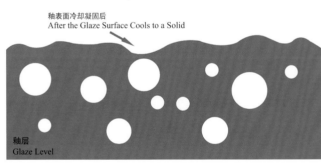

图6-3-44

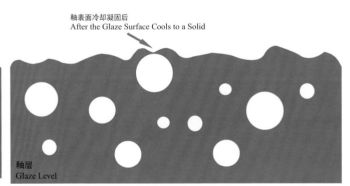

图6-3-45

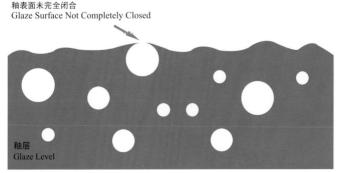

图6-3-46

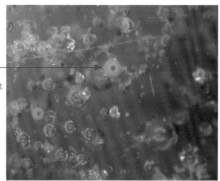
全变磨砂针眼型 无明显色阶变化
Fully-Changed Frosted Needlepoint Bubble, No Obvious Color Spectrum

图6-3-47（T30）

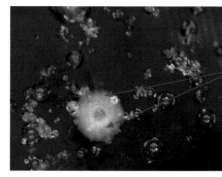
全变磨砂针眼型 有色阶变化
Fully-Changed Frosted Needlepoint Bubble with Color Spectrum

图6-3-48（B2）

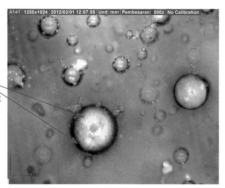
全变磨砂针眼型 有色阶变化
Fully-Changed Frosted Needlepoint Bubble with Color Spectrum

图6-3-49（J3）

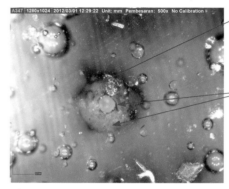
可见玻璃相
Glass Phase Visible

局变磨砂针眼型 有色阶变化
Semi-Changed Frosted Needlepoint Bubble with Color Spectrum

图6-3-50（J7）

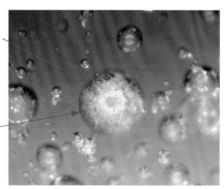
可见玻璃相
Glass Phase Visible

局变磨砂针眼与颗粒针眼并存
Semi-Changed Frosted Needlepoint and Grainy Needlepoint Bubble

图6-3-51（N12）

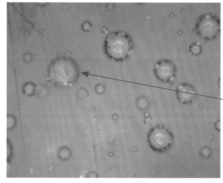
全变磨砂针眼型 无明显色阶变化
Fully-Changed Frosted Needlepoint Bubble, No Obvious Color Spectrum

图6-3-52（T2）

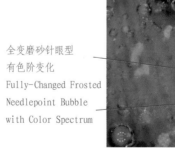
全变磨砂针眼型 有色阶变化
Fully-Changed Frosted Needlepoint Bubble with Color Spectrum

图6-3-53（T21）

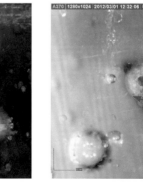
局变颗粒针眼型 有色阶变化
Semi-Changed Grainy Needlepoint Bubble with Color Spectrum

可见玻璃相
Glass Phase Visible

图6-3-54（J8）

B. 非圆口破口壁变泡

a. 非圆口破口壁变泡概念

非圆口破口壁变泡是破口为不规则形状的壁变泡。

b. 非圆口破口壁变泡特征

大部分非圆口破口壁变泡，尽管气泡破口不圆，但破口边沿仍是平滑的，无齿状感。通常有齿状边沿的破口泡，应该是新破不久的气泡。

非圆口破口壁变泡气泡的内壁的风化形貌与非破口壁变泡风化形貌类似（参见本章本节非破口壁变泡种类）；有磨砂针眼型（见图6-3-52、图6-3-53），也有颗粒针眼型（见图6-3-54—图6-3-57）。

有的非圆口破口壁变泡为全变（见图6-3-52、图6-3-53）；有的为局变，透过气泡还能见玻璃相（见图6-3-54、图6-3-56、图6-3-57）。

有的气泡破口下方可见物质存在，不可见玻璃相（见图6-3-52、图6-3-54）。有的气泡内壁有色阶变化（见图6-3-53、图6-3-55、图6-3-57），有的没

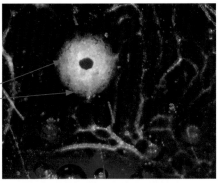

全变颗粒针眼型
有色阶变化
Fully-Changed Grainy Needlepoint Bubble with Color Spectrum

图6-3-55（T3）

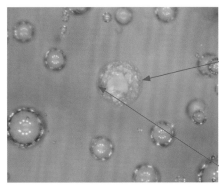

局变颗粒针眼型
无明显色阶变化
Semi-Changed Grainy Needlepoint Bubble, No Obvious Color Spectrum

可见玻璃相
Glass Phase Visible

图6-3-56（H1）

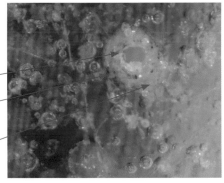

局变颗粒针眼型
有色阶变化
Semi-Changed Grainy Needlepoint Bubble with Color Spectrum
可见玻璃相
Glass Phase Visible

图6-3-57（T18）

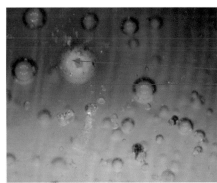

上浓下淡
Top Dark, Bottom Light

图6-3-58（T3）

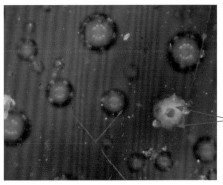

上浓下淡
Top Dark, Bottom Light

图6-3-59（T3）

有明显色阶变化（见图6-3-52、图6-3-56）。

C. 小结

在同一元青花瓷器上，圆口破口壁变泡与非圆口破口壁变泡可以同时存在。它们中大部分破口边沿处是平滑的，破口平滑率越高越能体现迹型的量变特征。

根据标本观察，部分破口壁变泡有色阶变化，无论是被物质全部充满还是部分充满，颜色顺破口处渐变，上浓下淡，这是因为与外界连贯的通道大都是先从气泡顶部连通的（见图6-3-58、6-3-59），也有的颜色差异不明显。

见图6-3-60、图6-3-61（图6-3-60、图6-3-61为同一器物同一位置图片，图6-3-60为二维图片，图6-3-61为三维图片）。

图6-3-60中的气泡M，在图6-3-61中位于红圈釉下处，不可见。图6-3-60箭头N所指破口，在图6-3-61中更清晰可见。破口周围釉表面较干净，且无明显变色。

由此可推断，气泡M内壁有风化形成的斑块。箭头N所指的破口，属于釉表面的结构缺陷，应是外界连接气泡内的通道。外界物质通过釉表面的通道进入釉层中的气泡M内，使气泡内壁发生质地变化，失去了透影性，形成壁变泡。

这些现象都说明，泡内风化现象是侵蚀物紧贴气泡的内壁，以积累的模式渐进风化，而不是弥漫在空间随机风化。气泡破口周围是颜色最早变化和颜色较深的地方，因为破口是气泡连接外界的通道。

（3）小结

在同一元青花瓷器上，非破口壁变泡与破口壁变泡可以同时存在。非破口泡内壁的风化程度一般要比破口壁变泡低，体现了风化的时间过程性。

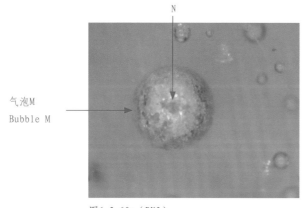

图6-3-60*（DY3）

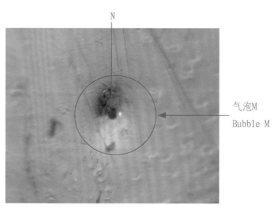

图6-3-61*（DY3）

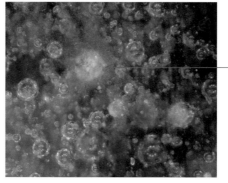

图6-3-62（DF1）

*图片经过裁剪放大处理。
*Picture cropped and enlarged to show detail.

## （三）假象壁变泡

### 1. 假象壁变泡概念

假象壁变泡是釉表面由于自然风化，或人为做旧形成的针眼斑、坑点斑、物盖斑等，恰好处于釉层中某一气泡的正上方，观察时易被误认为是壁变泡的气泡。

### 2. 假象壁变泡辨别方法

辨别假象壁变泡的方法有三种，它们分别是：直径比较法、焦距调整法、马士宁成像法。

当位于气泡上方斑块的最大直径超出气泡最大直径时，我们可以运用直径比较法和马士宁成像法来辨别；当位于气泡上方的斑块小于气泡直径时，我们可以运用焦距调整法和马士宁成像法来辨别。

（1）当斑块最大直径大于气泡最大直径时

运用直径比较法，当位于气泡上方斑块的最大直径超出气泡最大直径，且斑块将气泡完全覆盖时，可以直接判定该气泡属于假象壁变泡（见图6-3-62中气泡O）。

当位于气泡上方斑块直径大于气泡直径，但又没有将其完全覆盖时，我们需要采用马士宁成像法进一步判断。

见图6-3-63、6-3-64（图6-3-63、图6-3-64为同一器物同一位置图片，图6-3-63为二维图片，6-3-64为三维图片）。

图6-3-63中的气泡P，在图6-3-64中不可见，但其实质存在于箭头Q所指红圈釉下处；图6-3-63中箭头Q指出的深色斑块，在图6-3-64中也可见，其形状大小与图6-3-63相当。

图6-3-63中的气泡P是真正的非破口壁变泡，因为气泡的斑块都处于气泡内壁而非釉表面。两图中箭头Q所指深色斑块是在同一气泡上的同一位置，说明箭头Q所指斑块处于釉表面，属于结构有缺陷，应是外界连接气泡内的通道。

图6-3-64中除箭头Q所指深色斑块较明显外，周围釉表面基本光滑，这是壁变泡在漫长时间过程中的量变形成的反映。外界物质通过釉表面的通道不断进入釉层中的气泡内，使气泡内壁发生质地变化，失去了透影性。

（2）当斑块最大直径小于气泡最大直径时

当位于气泡上方斑块的最大直径小于气泡最大直径时，可以运用焦距调整法去判断。

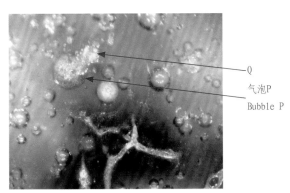

图6-3-63（DY7）

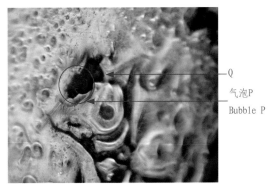

图6-3-64（DY7）

当摄像头调好焦距后，焦距对准的那一平面上的观察物都是最清晰的，凡不在同一平面上的其他物质成像都是模糊的。如果发现两个或多个斑块在调好焦距后观察仍有清晰与模糊的差别，说明有两种可能：

第一种，斑块一部分在泡内壁，一部分在釉表面；

第二种，斑块都处于泡内壁，但不在同一平面上。

焦距调整法一般只适用于体积较大的气泡，如果气泡太小，很难分辨出焦距调整前后物像清晰度的区别。这时，我们可以采用马士宁成像法协助判断。

见图6-3-65，由于气泡体积较小，使用焦距调整法无法分辨，需要采用马士宁成像法。（图6-3-65、图6-3-66为同一器物同一位置图片，图6-3-65为二维图片，图6-3-66为三维图片。）

图6-3-65中的斑块R，在图6-3-66中的釉表面清晰可见，也就是说，存在于釉表面的斑块R，造成了气泡"壁变"的假象，所以该气泡属于假象壁变泡。

见图6-3-67和图6-3-68（图6-3-67、图6-3-68为同一器物同一位置图片，图6-3-67为二维图片，图6-3-68为三维图片）。

图6-3-67中的气泡S，在图6-3-68中位于红圈釉下处，不可见。图6-3-67箭头U所指深色斑块，在图6-3-68中却消失了。图6-3-67中箭头T所指深色斑块，在图6-3-68中也可见。

由此可推断，箭头U所指斑块不在釉表面，而在气泡S内壁，是内壁风化形成的斑块。箭头T所指斑块处于釉表面，是釉表面风化形成的斑块，属于结构有缺陷，应是外界连接气泡内的通道。外界物质通过釉表面的通道进入釉层中的气泡内，使气泡内壁发生质地变化，失去了透影性，形成壁变泡。

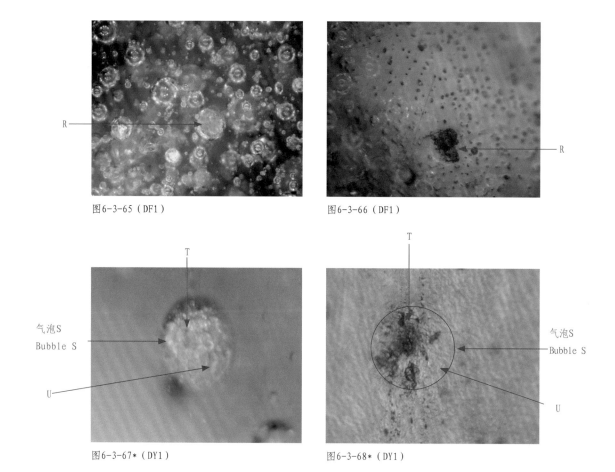

图6-3-65（DF1）　　　　　图6-3-66（DF1）

图6-3-67*（DY1）　　　　图6-3-68*（DY1）

*图片经过裁剪放大处理。
*Picture cropped and enlarged to show detail.

## 第四节 总结

陶瓷迹型包括在生产烧制过程中形成的先天性迹型，也有后天在环境变化中形成的后天性迹型。陶瓷迹型既有风化作用强度快与慢的相对性，也有风化强度高与低的绝对性。

不同朝代不同种类的陶瓷，由于用料和工艺的不同，会表现出不同的迹型。而陶瓷后天存放环境的不同，会使迹型具有复杂性和多样性。但是，特定朝代特定种类的陶瓷，总会有其普遍性特征。观察分析在绝对风化时间中迹型的普遍性特征，可以帮助我们判断特定朝代特定种类的器物。

本章列举了国内外元青花瓷器标本迹型。仔细观察分析，可以认识理解元青花迹型的普遍性特征。

元青花迹型在时间的量变过程中形成的各种特征，是图谱比对鉴定时的重要物证，深刻理解这些特征对元青花显微迹型的鉴定有着重要意义。

## Section 4:
# Summary

Ceramics Trace Model Study encompasses the innate traces formed during the firing of ceramics as well as those traces acquired under the influences of the post-firing environments. Ceramic traces are defined by the variations in fast or slow weathering intensities as well as the strong or weak absolute weathering severities.

Further, according to differences in materials and techniques, ceramics of different time periods or different types will present with their own set of characteristics. Similarly, variations in post-firing environments will ultimately bring forth complexity and multiplicity within the traces. However, each period and type of porcelain will all have their own common characteristics. The observation and analysis of these common absolute weathering time trace characteristics can help us to determine the specific age and specific type of a ceramic ware.

This chapter utilized YUB porcelain pieces, viewed both domestically and abroad as its sample base. Through meticulous observation and analysis, we can recognize and understand the characteristic traces common to YUB porcelain.

The quantitative change brought through the passage of time creates distinct characteristics in the traces of YUB porcelain. These are the instrumental material evidences that Standard Trace Atlas Model comparative authentication utilizes. An intimate understanding of these characteristics holds utmost importance in the authentication of YUB porcelain.

# 第七章
# 陶瓷迹型规律

**阅读提示：**

　　本书第四章、第五章阐明了陶瓷釉层的成分构成以及陶瓷釉层风化演变的过程，这为研究陶瓷迹型现象及规律提供了理论支撑。保存至今的古陶瓷上的诸多迹型，是古陶瓷经风化后表现出来的一系列有规律的现象。

　　规律是事物本质的必然联系。这种必然联系的特征是稳定和可重复的。陶瓷迹型规律实质是陶瓷迹型在风化过程中所表现出的一系列稳定、可重复的现象特征，这些现象特征是大自然的雕凿与造化，也是判断陶瓷真假的物证证据。本章以元代青花瓷器迹型为例，阐述了陶瓷迹型中的七大规律。这些规律是鉴定陶瓷所用的逻辑推理和逻辑论证的客观基础与依据。深刻理解这些规律对于陶瓷的断真识假有非常重要的指导意义。

**主要论述问题：**

　　1. 为什么说陶瓷迹型量变规律是陶瓷釉层自然风化的核心规律？

　　2. 陶瓷迹型的各种规律是如何在迹型与迹型之间的关系中表现出来的？

# Chapter 7
# Ceramic Trace Laws

Text Note:

Chapters 4 and 5 detailed ceramic glaze level composition and the processes of change through weathering. They provide theoretical support for the research of ceramic trace appearances and patterns. Ceramic trace phenomena are the series of laws that govern the appearances present in weathered ceramics.

Laws govern the inevitable relationships between the intrinsic characteristics of things. These relationships are characteristically stable and repeatable. Ceramic Trace Laws are essentially the explanation of the series of stable and repeating characteristics that manifest on weathered ceramic glazes. These characteristics are engraved and forged by nature, and act as the definitive material evidence for ceramic authentication. This chapter, using YUB porcelain as an example, lays out the Seven Ceramic Trace Laws. These laws are the objective foundation fundamental to the logical reasoning and proof finding in article authentication. Deeply understanding these laws is a particularly meaningful step in the authentication of ceramics.

# 第一节
# 陶瓷迹型随机性规律

## 一、陶瓷迹型随机性规律概念

陶瓷迹型随机性规律是在烧制过程中和后天环境影响下形成的，分布在釉表面及釉层中的陶瓷迹型具有不可预测性、非人为设定性的现象。

## 二、陶瓷迹型随机性规律成因

### （一）内因

内因是陶瓷本身的原因：
1. 陶瓷釉料中各种物质的构成及分布是随机的。
2. 釉层迹型类别在烧制完毕后的分布是随机的。
3. 釉层各处结构稳定性是随机的。

以上三点都是经风化作用后陶瓷迹型随机分布的前提。

### （二）外因

外因是外界环境变化影响的原因：

例如，两个相同的元青花梅瓶分别窖藏在两个密封缸中。其中一个密封缸破了，涌进了泥水；另一个缸密封则完好无损，十分干燥。那么两个梅瓶同时出土后，其釉层迹型随机分布的痕迹，肯定是不一样的。在涌进泥水环境中的那个梅瓶的迹型随机分布，要比在密封缸内处于干燥环境下的梅瓶复杂。干燥的那个梅瓶，是在一个相对等量风化的环境中度过的，而另一只却是在等量风化的环境中又增加了各种其他性质的增量风化，两只梅瓶的风化作用强度不一致。

陶瓷生存环境的不同以及环境变化的不可预测性导致风化相对时间的出现，形成了风化强度的随机性。

同样，在同一梅瓶上，不同位置的釉层在相对时间内的风化作用强度不同，其风化强度也表现出随机性。

## 三、陶瓷迹型随机性规律表现特征

陶瓷迹型随机性规律的表现特征是：陶瓷釉层迹型会随意出现在不同部位。

先天形成的各种陶瓷迹型种类分布随机，比如含铁氧化物的放射斑，可以出现在任何一处青料上，如出现在同一只碗的内部（见图7-1-1、图7-1-2，图7-1-2为图7-1-1中碗内部某处显微图片）和外部（见图7-1-3、图7-1-4，图7-1-4为图7-1-3中碗外部某处显微图片）。

图7-1-1（T19）

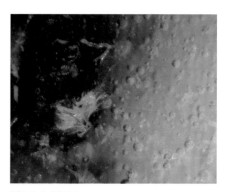

图7-1-2（T19）

图7-1-3（T19）

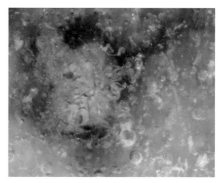

图7-1-4（T19）

## Section 1:
# Ceramic Trace Law of Randomness

## 1. Ceramic Trace Law of Randomness Defined

The Ceramic Trace Law of Randomness: The distribution of trace appearances across the glaze surface and within the glaze level, created during firing and under the influence of the post-firing environment, is unpredictable and a non-human-controlled phenomenon.

## 2. Ceramic Trace Law of Randomness Causes

### A. Internal Causes

Internal causes refer to the innate characteristics of ceramics:

I. The composition and distribution of materials in ceramic glaze is random.

II. The distribution of ceramic trace categories directly after firing is random.

III. The areas of structural stability within the glaze level are random.

The above three reasons are the precursors to the randomness seen in post-firing weathering.

### B. External Causes

External causes refer to the influences of the external environment.

For Example: If two identical YUB vases are stored in two separate sealed jars. The first jar is broken, allowing dirty water to enter into the vessel, and the second jar remains completely intact and completely dry. When the vases inside the jars are excavated, the random distribution of glaze level traces will be completely different. The random distribution of traces on the vase within the muddy water environment will most likely show drastically more complexity than that of the protected, dry vase. The dry vase is in a relatively simple weathering environment, while the second has the introduction of additional weathering. The weathering intensities experienced by the two vases are varied.

The unpredictability of the environment leads to the phenomenon of relative weathering periods, and as such, randomness in weathering severity occurs.

Additionally, on different points of a single vase, the relative weathering intensities experienced are different. Therefore, differing weathering severities appear, and the results of weathering assume certain randomness.

## 3. Ceramic Trace Law of Randomness Characteristics

The Ceramic Trace Law of Randomness characteristics manifest arbitrarily across different portions of the ceramic glaze.

The distribution of each of the trace categories formed during firing occurs randomly. For example: The radiating mottles containing Iron Oxide can appear at any point on the ceramic glaze that has blue dye; for instance the interior (see pictures 7-1-1 and 7-1-2; with 7-1-2 as the microscopic image of the interior portion of the bowl as shown in picture 7-1-1) and exterior (see pictures 7-1-3 and 7-1-4; with 7-1-4 as the microscopic image of the external portion the bowl as shown in picture 7-1-3) of the same bowl.

The distribution of post-firing trace categories is also random: For example: Color change bubbles can occur on

后天环境影响下形成的各种陶瓷迹型种类分布也随机，比如色变泡的出现，既可以在青花纹饰处（见图7-1-5），也可以在白釉处（见图7-1-6）。

图7-1-7
草原上随机分布牲畜
Random Grazing Distribution

图7-1-5（DY7）

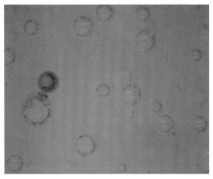

图7-1-6（DY3）

一同出土的同时期同类型的陶瓷，由于各自所处的环境变化的差异，从肉眼上看有的旧如新，有的痕迹斑驳，它们的迹型类别及种类特征（无论先天形成的还是后天形成的）都有很大差异。

## 四、小结

陶瓷迹型随机性规律说明了陶瓷迹型有不可预测和不可人为控制的特点，这既是陶瓷釉层本身结构稳定性的差异所致，也是后天生存环境的变化所致。

在微观世界与宏观世界中，都存在着随机性规律（见图7-1-7[103]）。

areas of blue dye (see picture 7-1-5) or areas of white glaze (see picture 7-1-6).

Additionally, the trace characteristics (whether innate or acquired) on excavated pieces of the same period and same type, appearing to the naked eye as either almost new or distinctively aged, show characteristics of categories and types that are drastically dissimilar.

## 4. Summary

The Law of Randomness shows that ceramic trace characteristics are unpredictable and uncontrollable, due to the inherent differences in glaze structure as well as differences in post-firing environments. Interestingly, the Law of Randomness in the microscopic and macroscopic realm is the same (see picture 7-1-7 [103]).

# 第二节 陶瓷迹型色差规律

## 一、陶瓷迹型色差规律概念

陶瓷迹型色差规律是在烧制过程中和后天环境影响下形成的，分布在釉表面及釉层中的同一陶瓷迹型种类表现出颜色差异的现象。

为了方便解释色差规律，在某一个迹型单项上发现的颜色差异称为色阶变化，在同一迹型种类中的多个迹型单项上发现的颜色差异称为色差变化。

比如，在同一个色变泡内的颜色差异叫色阶变化，两个或两个以上色变泡的颜色差异叫色差变化。

## 二、陶瓷迹型色差规律成因

就同一迹型单项而言，陶瓷迹型色阶的成因主要有：

1．各处釉层结构稳定性的差异。

2．各处风化强度的差异。

就同一迹型种类中的多个迹型单项而言，陶瓷迹型色差的成因主要有：

1．迹型单项所处釉层结构稳定性的差异。

2．迹型单项的风化强度的差异。

3．迹型单项的风化作用的差异。

## 三、陶瓷迹型色差规律表现特征

在不同陶瓷迹型类别的风化痕迹中，都可以发现它们有颜色的变化，这种颜色的变化既可以色阶方式在某一迹型单项上出现，也可以色差方式在同一迹型种类中的多个单项上出现。

以色阶方式在某一迹型单项上出现：

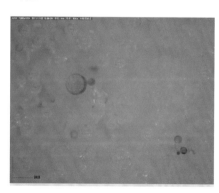

图7-2-1（DY7）

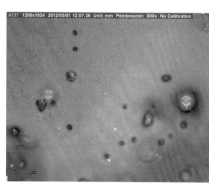

图7-2-2（J3）

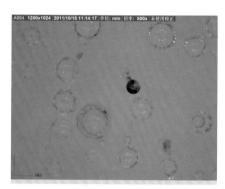

图7-2-3（DY1）

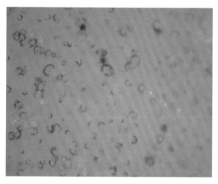

图7-2-4（S3）

如同一色变泡上的色阶变化（见图7-2-1）。

同一流淌斑上的色阶变化（见图7-2-2）。

以色差方式在同一迹型种类中的多个单项上出现：

如有色差变化的多个色变泡（见图7-2-3）。

有色差变化的C形、O形纹路群（见图7-2-4）。

## 四、小结

陶瓷迹型的颜色变化是细微与

# Section 2:
# Ceramic Trace Law of Chromatic Aberration

## 1. Ceramic Trace Law of Chromatic Aberration Defined

The Ceramic Trace Law of Chromatic Aberration states: Within a single trace type, distributed across the glaze surface and with the glaze level, both innate and acquired, the phenomenon of chromatic aberration (differences in color) can be present.

For ease of understanding the Law of Chromatic Aberration we say: If color variation within an individual trace occurs, we refer to this as a color spectrum change; if color variation within a trace type occurs, we call this color disparity change.

For example: If a single bubble presents with different colors, this is called color spectrum change; while if two or more bubbles present with different colors, this is a color disparity change.

## 2. Ceramic Trace Law of Chromatic Aberration Causes

Speaking of an individual trace, the main causes of color spectrum include:

Differences in the glaze structure stability at different points;

Different weathering severity.

Speaking of color differences between different traces within one trace type, the main causes of color disparity are:

Differences in the glaze structure stability surrounding each trace;

Differences in the weathering severity within each trace;

Differences in the weathering of each individual trace.

## 3. Ceramic Trace Law of Chromatic Aberration Characteristics

In the weathering traces left in each ceramic trace category, it is possible to see color changes. This chromatic aberration can assume a color spectrum within a single trace or appear as a color disparity within a trace type.

Color spectrum within a single trace:

A color spectrum presenting in a single color change bubble (see picture 7-2-1);

A color spectrum presenting on a single flowing mottle (see picture 7-2-2).

Color disparity within a single trace type:

A color disparity presenting in several color change bubbles (see picture 7-2-3);

A color disparity presenting amongst a grouping of C and O shaped striae (see picture 7-2-4).

## 4. Summary

Ceramic trace color change is fine and minute, with the chromatic aberration within each trace type showing that every area of the glaze can incur dissimilar weathering intensities, thus bringing about the phenomenon of higher and lower weathering severity.

渐进的，每种迹型种类的颜色变化，都反映了釉层各处风化作用强度不一致以及风化强度差异的现象。

# 第三节 陶瓷迹型量变规律

## 一、陶瓷迹型量变规律概念

陶瓷迹型量变规律是分布在陶瓷釉表面及釉层中的迹型在后天的风化过程中，有着缓慢性、渐变性、细腻性、精致性、分明性等特征的现象。

## 二、陶瓷迹型量变规律成因

古代陶瓷流传至今，都经过了漫长的时间。在这段时间里，不论保存环境如何，陶瓷釉层一定会发生风化作用。由于陶瓷釉层结构比较稳定，抗侵蚀能力较强，所以陶瓷釉层的风化作用是外界对其缓慢的物理堆积与化学溶蚀过程。

## 三、陶瓷迹型量变规律表现特征

以元青花瓷器的迹型为例，有如下多种特征：

### 1. 曲线性特征

如物盖斑的边界（见图7-3-1）；洗去物盖斑后，留下的针眼斑边界（见图7-3-2）。

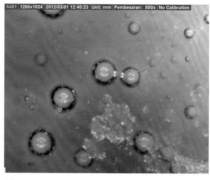

图7-3-1（J11）

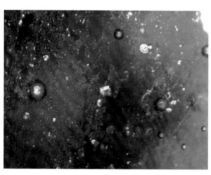

图7-3-2（B7）

### 2. 细腻性特征

如雾状针眼斑在沙滩斑上的分布，有从浓到淡的均匀过渡（见图7-3-3）。

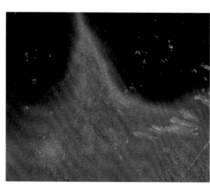

图7-3-3（T21）

### 3. 内敛形特征

如C形纹、O形纹，紧致而收敛，没有发散的毛边（见图7-3-4）。

分布在流淌斑形貌内的雾状针眼斑、磨砂针眼斑，几乎都是在流淌型形貌之内的，收敛不扩张（见图7-3-5）。

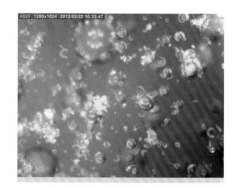

图7-3-4（Z1）

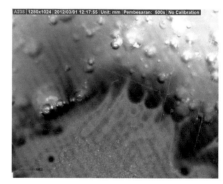

图7-3-5（J4）

### 4. 均匀性特征

雾状针眼斑在流淌斑上以及流淌纹上分布均匀（见图7-3-6、图7-3-7）。

Section 3:
# Ceramic Trace Law of Quantitative Change

## 1. Ceramic Trace Law of Quantitative Change Defined

The Ceramic Trace Law of Quantitative Change states: The innate and acquired trace appearances distributed across the glaze surface and within the glaze level, have all undergone slow, gradual, and minute processes and appear with fine and distinct characteristics.

## 2. Ceramic Trace Law of Quantitative Change Cause

Ancient ceramics that have been passed down to today have undergone a long passage of time. During this period, regardless of how well protected the piece is, the ceramic glaze has experienced a certain amount of weathering intensity. Due to the relative structural stability and corrosion-resistant nature of ceramic glaze, the process of weathering is the gradual amassing of physical change and chemical corrosion brought on by the external environment.

## 3. Ceramic Trace Law of Quantitative Change Characteristics

According to the quantitative change processes occurring on YUB porcelain, the Law of Quantitative Change characteristics are:

### A. Irregular Border Characteristics:

The border of overlay mottles for example (see picture 7-3-1), once overlay mottles are cleaned away, the remaining needlepoint mottle boundaries (see picture 7-3-2) resemble that of the irregular borders formed by coastlines.

### B. Minute Characteristics:

Foggy needlepoint mottling distribution for example (see picture 7-3-3), appear with a dark to light even progression.

### C. Restrained Characteristics:

C and O shaped striae edges are tightly constrained instead of roughly dispersed for example (see picture 7-3-4).

Foggy and frosted needlepoint mottling on flowing mottles almost entirely occurs on the interior of the appearance; it is tightly restrained not widely dispersed (see picture 7-3-5).

### D. Homogenous Characteristics:

Looking at the distribution of foggy needlepoint mottling on flowing mottles (see picture 7-3-6) and on flowing striae (see picture 7-3-7), all are homogenously distributed.

### E. Fine Characteristics:

The finely exquisite thin border of radiating mottles for example (see picture 7-3-8); also seen are very similar characteristics in nature, i.e. the thin boundary around islands created by ocean waves (see picture 7-3-9 [104]); during calm weather, the waves surrounding a small island creates a fine border, this is representative of quantitative change; however in times of perilous storms, the boundaries created by crashing waves are broad and coarse, this is representative of sudden change (see picture 7-3-10 [105]).

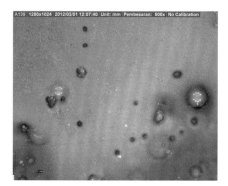

图7-3-6（J3）

图7-3-7（Z10）

### 5. 精致性特征

正如在宏观世界中,被海水包围的小岛:天气晴朗时海水在小岛周围形成纤细包边,是量变特征(见图7-3-9[104]);而狂风、大浪造成的包边是质变特征,是扩张的、不精致的(见图7-3-10[105])。

而放射斑周围的包边就像天气晴朗时海水在小岛周围形成的包边,纤细精致(见图7-3-8)。

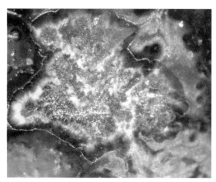

图7-3-8（N4）

图7-3-9
复活节岛上的拉拉库努火山
Easter Island

图7-3-10

### 6. 自然美特征

不同迹型类别表现出自然的美感。显微与宏观的特性,在大自然造化下,惊人相似,如出一辙。因为它们的物质基础——原子分子的构成都是相同的。

图7-3-11为用马士宁成像技术拍摄的放射斑,图7-3-12[106]为国际宇宙空间站宇航员拍摄的智利灰冰川。

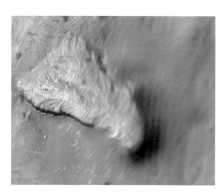

图7-3-11（B3）

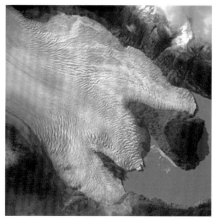

图7-3-12

图7-3-13为元青花标本上的放射斑,图7-3-14[107]为从太空拍摄的冰岛西北部海湾附近的冰川。

图7-3-13（B3）

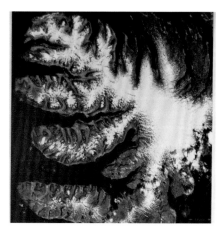

图7-3-14

图7-3-15为元青花标本上的流淌斑,图7-3-16[108]为飞机上拍摄的海中小岛及沙滩。

## F. Natural Beauty Characteristics:

There is an inherently exquisite natural beauty that appears in ceramic trace categories, very similar to the beauty seen macroscopically in nature. This is because the foundational elements of each are the same.

For Example:

Picture 7-3-11 is a picture taken of a radiating mottle using the Bunney Method, and picture 7-3-12 [106] is a picture taken by members of The International Space Station of Chile's Grey Glacier.

Picture 7-3-13 is an image of a radiating mottle on a YUB sample; 7-3-14 [107] is an aerial picture of a glacier on the Northeast coast of Iceland.

Picture 7-3-15 is an image of a flowing mottle on a YUB sample; 7-3-16 [108] is an aerial picture of an island and beach.

Picture 7-3-17 is an image of needlepoint mottling on a YUB sample, 7-3-18 [109] is an aerial picture, from 400 meters, of a flock of flamingos in the Lake Nakuru National Park in Kenya.

Picture 7-3-19 is an image of flowing striae on a YUB sample, 7-3-20 [110] is a picture of the random tracks left in wet soil by earthworms.

Picture 7-3-21 is an image of flowing striae on a YUB sample, 7-3-22 [111] is a photo taken during a lightning storm.

Picture 7-3-23 is an image taken from a Lang Kiln Red vase from the Stamen Collection, 7-3-24 [112] is Tycho's Supernova, researched in 1572 by Danish astrologer Tycho Brahe.

## 4. Summary

The Ceramic Trace Law of Quantitative Change brings to light the phenomenon of ceramic trace weathering; the slow and gradual development of change over a long period of time. The characteristics of this weathering development appear in each of the ceramic trace categories as: Round, Minute, Restrained, Homogenous, and Fine. These microscopic appearances are paralleled by the macroscopic beauty seen in nature; both occuring naturally.

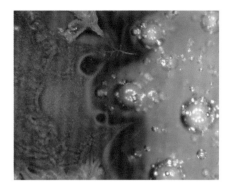

图7-3-15（H5）

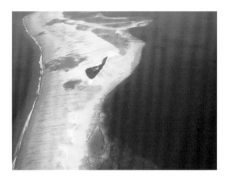

图7-3-16

图7-3-17为元青花标本上的针眼斑，图7-3-18为在飞行高度400米上空拍摄的栖息在肯尼亚纳库鲁湖国家公园的火烈鸟群[109]。

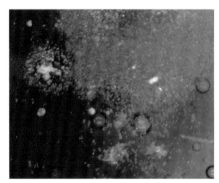

图7-3-17（T21）

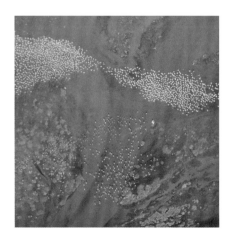

图7-3-18

图7-3-19为元青花标本上的流淌纹，图7-3-20[110]为蚯蚓在土壤中移动留下的痕迹。

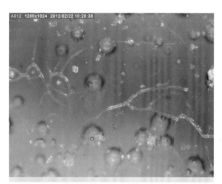

图7-3-19（Z1）

图7-3-20

图7-3-21为元青花标本上的流淌纹，图7-3-22[111]为大自然的闪电现象。

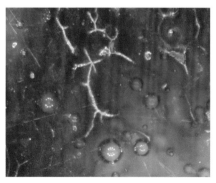

图7-3-21（N2）

图7-3-22

图7-3-23为美国洁蕊堂藏品清康熙郎窑红釉蒜头瓶上的破口壁变泡，图7-3-24[112]为1572年丹麦天文学家第谷·布拉赫研究的第谷超新星遗迹。

图7-3-23（JS1）

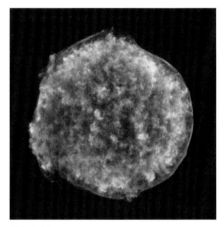

图7-3-24

## 四、小结

陶瓷迹型量变规律揭示出陶瓷迹型的风化过程是在时间长河中缓慢、渐进的发展。这种风化发展的特征，表现在各迹型类别的形貌特征上，是曲线的、细腻的、内敛的、均匀的、分明的、精致的。这些显微现象与大自然中许多宏观现象惊人相似，它们都是大自然鬼斧神工的杰作。

# 第四节
## 陶瓷迹型层次性规律

### 一、陶瓷迹型层次性规律概念

陶瓷迹型层次性规律是陶瓷迹型不同的类别之间和同一类别之间风化强度的差异，通过它们在釉层（釉表面或釉层中）纵向垂直的位置表现出来的现象。

### 二、陶瓷迹型层次性规律成因

陶瓷迹型层次性规律的成因主要有三方面：

1．就某一迹型单项而言，由于其所处釉层不同位置的结构稳定性有一定差异，或者该迹型单项受到釉层保护的程度不同，从而影响其风化强度高低。

一般来说，处于釉表面的迹型单项相对于釉层中的迹型单项，风化强度更高。例如，处于釉表面的流淌斑无釉层保护，风化强度高；受到釉层保护的流淌斑风化强度低。风化强度的差异通过流淌斑的层次性体现出来。

2．就相邻的两个或多个迹型单项（都属同一迹型类别）而言，风化强度有差异，会表现出层次性。如坑点斑下的絮状斑。

3．就相邻的两个或多个迹型单项（属不同迹型类别）而言，风化强度有差异，也会表现出层次性。如C形、O形纹下的坑点斑。

### 三、陶瓷迹型层次性规律特征

陶瓷迹型层次性规律是陶瓷迹型不同类别或同一类别的风化强度差异，在釉层中纵向垂直的位置表现出来的不同特征。

1．就某一迹型单项而言，流淌斑分布于釉表面及釉层中，受釉层保护的程度不一，这使它们各处风化强度有差异。针眼斑在流淌斑上疏密不一的分布，就是流淌斑风化强度不一的表现。流淌斑的分布特征以及覆盖在上面的针眼斑分布特征使其有了层次感（见图7-4-1），这与大自然中的海浪和沙滩呈现出的层次感类似。

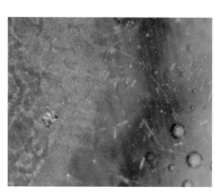

图7-4-1（T3）

2．就同一类别中相邻的两个或多个迹型单项而言，坑点斑形成后成为连接外界物质与釉层内部的通道之一，长期存放于干湿交替的环境中，釉下絮状斑会在坑点斑下的釉层中及坑点斑周围的釉层中逐渐形成，坑点斑与釉层中存在的絮状斑就有了层次关系（见图7-4-2）。

3．就不同类别中相邻的两个或多个迹型单项而言，由C形、O形纹发展而来的坑点斑，这两种不同类别之间的迹型种类也可表现出层次性。图7-4-3表现了从C形、O形纹发展成的坑点斑的时间过程，从存在于釉表面的C形、O形纹发展到深入釉层中的坑点斑，由浅到深，富有层次。

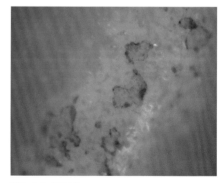

图7-4-2（T31）

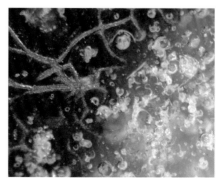

图7-4-3（T36）

# Section 4:
# Ceramic Trace Law of Gradation

## 1. Ceramic Trace Law of Gradation Defined

The Ceramic Trace Law of Gradation (Layering) states: The weathering severity, whether inter or intra-trace category has variation; a phenomenon resulting due to their positioning throughout the glaze level (across the glaze surface or within the glaze level).

## 2. Ceramic Trace Law of Gradation Causes

There are three primary aspects to the causes of the Ceramic Trace Law of Gradation:

A. In reference to a single trace, the differences in structural stability at various points of the glaze determine the amount of protection offered to each trace, and affect the weathering severity.

In general, traces on the glaze surface receive a higher intensity of weathering than traces within the glaze level. For example: Flowing mottles on the glaze surface receive less protection from the glaze, and therefore have a much higher severity of weathering, and mottles within the glaze level receive better protection from the glaze, hence a relatively low severity; creating gradation in weathering.

B. In reference to two or more individual traces (belonging to the same category), due to varying weathering severity, gradation can present: For example, floccus mottling under pitting.

C. In reference to two or more individual traces (belonging to different categories), due to varying weathering severity, gradation can present. For example, pitting occurring under C and O shaped striae.

## 3. Ceramic Trace Law of Gradation Characteristics

The Ceramic Trace Law of Gradation phenomenon characteristics appear differently among different trace categories and within a trace category according to differences in weathering intensity at different points. For Example:

A. In reference to an individual trace, flowing mottles are located above and below the surface of the glaze, and receive different degrees of glaze protection; allowing a range of weathering severity. Therefore, the differences in density of needlepoint mottling are a direct result of differences in weathering severity. The characteristics of flowing mottles and associated overlaid needlepoint mottling appear with a feeling of layers or gradation (see picture 7-4-1). This is similar to the gradation feel observed in the natural relationship between waves and the beach.

B. In reference to adjacent traces of the same category, underglaze floccus mottling forms via the glaze level structural flaws such as pitting; bringing rise to gradation within a single trace category: The pit mottle formed a channel between the glaze level and the external environment, allowing for the gradual formation of floccus mottling. The relationship between the pitting and floccus mottling is one of gradation (see picture 7-4-2).

C. In reference to adjacent traces of different categories, the progression from C and O shaped striae into pit mottles represents the layered relationship between different trace categories. Picture 7-4-3 shows the temporal process of progression from C to O shaped striae to pit mottles; varying depths giving a feeling of gradation.

## 四、小结

陶瓷釉层在长期风化过程中，陶瓷釉层迹型呈现出由表及里、由此及彼的层次性。

陶瓷迹型层次性规律强调纵向观察与比较——釉表面至釉层中的风化强度的差异现象。

# 4. Summary

According to the long weathering process and differences in weathering severity at different points of the glaze level and surface, gradation appears.

The Ceramic Trace Law of Gradation emphasizes the vertical observation and comparison of all weathering severities within the glaze level and across the glaze surface.

# 第五节 陶瓷迹型多样性规律

## 一、陶瓷迹型多样性规律概念

陶瓷迹型多样性规律是陶瓷在后天环境影响下，在釉表面及釉层中，因风化而形成的迹型，比其先天形成的迹型类别或种类数量增加的现象。

## 二、陶瓷迹型多样性规律成因

陶瓷迹型多样性规律成因主要是陶瓷存放环境的不断变化，相对风化时间不断增多，导致更多的迹型类别或种类出现。

## 三、陶瓷迹型多样性规律特征

同一时期同一类型的陶瓷，因存放的环境不同或相同环境中的局部环境发生了变化，其迹型特征表现会有很大差异。这种差异表现在迹型类别及种类的多样性上。生存环境好，环境变化不大，风化作用强度差异不大，风化强度较低的陶瓷，其迹型有简单性特征，比如，釉表面比较干净清晰，迹型类别和种类变化都不明显或增加迹型类别和种类数量较少（见图7-5-1、图7-5-2）。

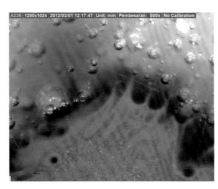

图7-5-2（J4）

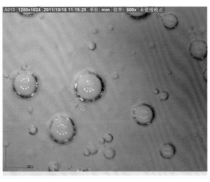

图7-5-1（DY1）

而生存环境不好，环境变化大，风化作用强度差异大，且风化强度高的陶瓷，其迹型呈多样性特征，会有坑点斑、划痕、壁变泡等多种迹型种类风化严重的现象出现（见图7-5-3、图7-5-4）。

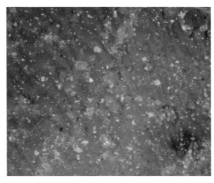

图7-5-3（T33）

图7-5-4（T2）

通过一些考古发掘的器物可知，无论从水中，还是从地下发掘出来的古陶瓷，在同一批次中，会发现有的陶瓷风化强度较高，有的较低。如法门寺地宫出土的秘色瓷，有的光亮如新，有的老旧斑驳，有时甚至用肉眼就能看出它们有"新旧区别"。

## 四、小结

在陶瓷迹型规律中，迹型多样性规律是陶瓷保存环境的复杂性及时间漫长性的反映，是古陶瓷迹型普遍的特征。陶瓷存放环境恶劣，且变化大，导致风化作用强度的差异大，风化强度相对高，其迹型多样性就表现突出。而陶瓷迹型的简单性，是陶瓷存放环境较好，环境变化小的反映，属迹型特殊性特征。不能因为某些陶瓷迹型现象相对简单而武断地将其判定为假。

# Section 5:
# Ceramic Trace Law of Variegation

## 1. Ceramic Trace Law of Variegation Defined

The Ceramic Trace Law of Variegation describes the phenomenon of an increase in appearances of trace categories and types distributed across the glaze surface and throughout the glaze level due to weathering incurred in the post-firing environment.

## 2. Ceramic Trace Law of Variegation Cause

The phenomenon of ceramic trace variegation is caused by the constant changes to the post-firing environment, thus the continual increase of variable weathering time periods, leading to the appearance of more and more trace categories or types.

## 3. Ceramic Trace Law of Variegation Characteristics

The ceramic trace characteristics on ceramic pieces of identical period and classification can appear differently. These dissimilarities are shown in the variegation within trace categories and trace types. Ceramics from relatively simple environments with only small differences in weathering intensities appear with relatively simple characteristics and low weathering severity. For example, the glaze surface may be basically clean and clear, trace category and type changes are not obvious, or a comparatively low number of changed trace categories and types exist (see pictures 7-5-1 and 7-5-2).

Ceramics from relatively complex environments and comparatively intense or more severe weathering show quite diverse trace change characteristics: Pitting, scrapes, and textured bubbles. For example are varieties of trace changes indicating higher weathering severity (see pictures 7-5-3 and 7-5-4).

Archeological excavations indicate: Regardless of whether excavated from water or earth, ancient ceramics within a single batch can appear with some pieces experiencing weak weathering and others very strong. Portions of the 'secret' color porcelain excavated from the underground palace of Famen Temple look almost new, and some are notably old and motley. Even using the naked eye, you can clearly differentiate the 'old and new'.

## 4. Summary

In Ceramic Trace Laws, trace variegation is the universal reflection of the complex environment and the passage of time in weathering of ancient ceramics: Ceramics stored in harsh and vastly changing environments experience widely diverse weathering intensities; the weathering severity is comparatively high and trace variegation is clear. On the other hand, trace simplicity is the reflection of a simple, scarcely changed environment, but is characteristically atypical. However, we can not simply view the relatively simplistic nature of the traces as a determining factor that an item is a forgery.

## 第六节
# 陶瓷迹型差异性规律

## 一、陶瓷迹型差异性规律概念

陶瓷迹型差异性规律是，陶瓷迹型类别或迹型种类的风化强度，与其邻近的相同或不同迹型类别、迹型种类不一致的现象。

## 二、陶瓷迹型差异性规律成因

陶瓷迹型差异性规律成因主要是：迹型类别及种类的成分不同，它们各自在釉层所处的位置结构稳定性不同，以及在相对时间内，各自风化作用强度的差异，形成了风化强度高低共存的现象。

## 三、陶瓷迹型差异性规律特征

当某一迹型类别的风化强度较高时，与其相邻的相同或不同的迹型类别却呈现出风化强度较低的特征时，陶瓷迹型差异性规律就表现出来了。如：破口壁变泡周围是干净清晰的玻璃相（见图7-6-1）；螺旋叠加的坑点斑周围是干净清晰的玻璃相等（见图7-6-2）。

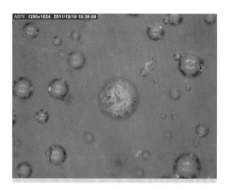

图7-6-1（DY3）

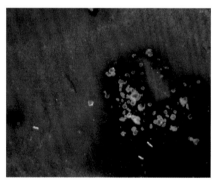

图7-6-2（T21）

## 四、小结

陶瓷迹型差异性规律揭示了陶瓷釉层迹型在后天环境中，风化强度不一致的特征。不同的迹型类别和种类之间风化强度的差异越大，陶瓷迹型差异性特征越明显。

陶瓷迹型差异性规律强调：横向观察与比较——釉表面至釉层中的风化强度的差异现象。

# Section 6:
# Ceramic Trace Law of Disparity

## 1. Ceramic Trace Law of Disparity Defined

The Ceramic Trace Law of Disparity describes the phenomenon of two similar or dissimilar ceramic trace categories or types within the same vicinity appearing with differences in weathering severity.

## 2. Ceramic Trace Law of Disparity Cause

The main cause of the Ceramic Trace Law of Disparity is: The composition of trace categories and types is different. Also, the respective locations and associated structural stability of the surrounding glaze, as well as variations in weathering intensities during different variable weathering times, all act to bring about the phenomenon of the coexistence of wide-ranging weathering severities.

## 3. Ceramic Trace Law of Disparity Characteristics

When the severity of weathering of a certain trace category is relatively high, and the weathering of adjacent trace categories is characteristically low, the Ceramic Trace Law of Disparity manifests. For example: The clean and clear glass phase surrounding a broken mouthed textured bubble (see picture 7-6-1); the clean and clear glass phase surrounding the spiral helix pitting (see picture 7-6-2).

## 4. Summary

The Ceramic Trace Law of Disparity reveals the effects brought on by the differences in weathering severity of ceramic glaze level traces. The larger the differences in weathering severity between trace categories and types, the larger the disparities of trace category and type appearances, and the more distinct the dissimilarities created by the passage of time become.

The Ceramics Trace Law of Disparity emphasizes the horizontal observation and comparison of all weathering severities across the glaze surface and within the glaze level.

# 第七节
# 陶瓷迹型分明性规律

## 一、陶瓷迹型分明性规律概念

陶瓷迹型分明性规律，是陶瓷在后天环境影响下形成的，分布在釉表面及釉层中的迹型类别或种类之间保持界限清晰，各自独立的现象。

## 二、陶瓷迹型分明性规律成因

陶瓷迹型分明性规律成因主要是：

1. 各迹型类别及种类的成分不同，在风化时会发生不一样的物理变化和化学反应，这些变化与反应是独立进行的，所以形成的迹型特征是独立的且形貌各异。

2. 迹型类别及种类各自在釉层所处的位置结构稳定性不同，因而形成风化强度的差异。

3. 陶瓷的存放环境发生变化，陶瓷在相对时间内受到随机的风化侵扰，陶瓷各部位的风化作用强度有差异，因而形成风化强度的差异。

## 三、陶瓷迹型分明性规律特征

不同的迹型类别或种类，其各自的边界大都是清晰的：

如：流淌纹与周围玻璃相的界线清晰分明（见图7-7-1）；放射斑与其周围的流淌斑之间界线清晰分明（见图7-7-2）；壁变泡与其周围的玻璃相界线清晰分明（见图7-7-3）。

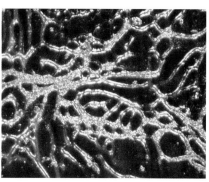

图7-7-1（DY3）

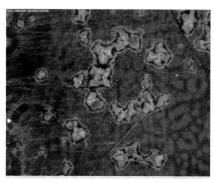

图7-7-2（DY1）

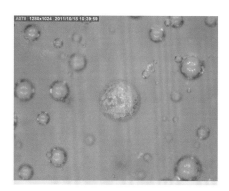

图7-7-3（DY3）

## 四、小结

陶瓷迹型分明性规律揭示了陶瓷釉层在后天环境影响下，各迹型类别与种类分布的独立性特征。这种独立性特征是量变特征的体现。

Section 7:
# Ceramic Trace Law of Distinction

## 1. Ceramic Trace Law of Distinction Defined

The Ceramic Trace Law of Distinction describes the phenomenon of trace categories and types distributed across the glaze surface and within the glaze level, formed within the post-firing environment, retaining clear delineation and respective independence from one another.

## 2. Ceramic Trace Law of Distinction Causes

The three main causes of the Ceramic Trace Law of Distinction are:

A. Different compositions of trace categories and types, as well as the dissimilarities of independently occurring physical changes and chemical reactions, leads to the independent and unique formation of trace characteristics;

B. As the glaze structural stability is different at each point of a trace category or type, the weathering severities and outcomes are correspondingly different;

C. The external environment of a ceramic piece undergoes change, the item undergoes variable weathering corrosion within each variable time period, and associated weathering intensities are different across each portion of the ware; thus bringing about divisions in weathering severities.

## 3. Ceramic Trace Law of Distinction Characteristics

Through the research and observation of YUB porcelain, we have found that regardless of weathering severity to trace categories and types, they maintain clearly delineated boundaries. For example:

The boundary between flowing striae and the surrounding glass phase is clearly distinct (see picture 7-7-1).

The boundary between radiating and flowing mottles is clearly distinct (see picture 7-7-2).

The boundary between textured bubbles and the surrounding glass phase is clearly distinct (see picture 7-7-3).

## 4. Summary

The Ceramic Trace Law of Distinction describes the independent nature of trace categories and types that were formed under the influence of the post-firing environment. This independence is characterized by clear, distinct boundaries between trace categories or trace types. This independence is the characteristic manifestation of quantitative change.

## 第八节
# 总 结

　　陶瓷迹型的七大规律是通过对陶瓷迹型的形成、演变过程以及它们之间的关系，从不同角度、不同方面进行观察、分析及研究，最终整理、归纳、揭示出的一套规律。

　　这套规律是以陶瓷迹型在漫长的量变过程中的变化特征为基础，从不同角度揭示了陶瓷迹型之间的客观必然联系。这些必然联系是对陶瓷的年代及窑口进行鉴定时，所用的逻辑推理和逻辑论证的客观基础与依据。它能帮助我们理解自然风化的现象与人工做旧现象的本质区别。

## Section 8:
# Conclusion

The Seven Ceramic Trace Laws describe the formation and development of ceramic trace categories and types. Only through the use of multiple angles and methods of observation, strict analysis, and laborious research; were we able to ultimately surmise and arrange this set of laws.

This set of laws is based on the premise that ceramic traces have undergone quantitative changes over the passage of a long period of time; revealing from different angles the objective positive correlations among ceramic traces. These positive connections are the objective foundations and fundamentals on which the logical inferences and proofs for determining the period and kiln of a ceramic ware are based; assisting us to further understand the critical differences between the natural change phenomena and the forced changes brought on by man-made aging techniques.

# 附录 Appendix

本书图片选用的陶瓷标本，除代号为DF1的瓷片为现代仿品，其余均为真品。

The pictures used in this text are of ceramic samples. Of these, piece marked DF1 is a modern forgery, the remaining are all authentic.

| 代号<br># | 来源单位<br>Source | 标本图片<br>Picture | 显微图片<br>Microscopic Image | | | |
|---|---|---|---|---|---|---|
| DY1 | 广州东方博物馆<br>Guangzhou Oriental Museum | | | | | |

(续表)

| 代号 # | 来源单位 Source | 标本图片 Picture | 显微图片 Microscopic Image | | | |
|---|---|---|---|---|---|---|
| DY3 | 广州东方博物馆 Guangzhou Oriental Museum | | | | | |

(续表)

| 代号<br>#  | 来源单位<br>Source | 标本图片<br>Picture | 显微图片<br>Microscopic Image ||||
|---|---|---|---|---|---|---|
| DY5 | 广州东方博物馆<br>Guangzhou Oriental Museum | | | | | |
| DY7 | 广州东方博物馆<br>Guangzhou Oriental Museum | | | | | |

(续表)

| 代号 # | 来源单位 Source | 标本图片 Picture | 显微图片 Microscopic Image | | |
|---|---|---|---|---|---|
| DQ1 | 广州东方博物馆 Guangzhou Oriental Museum | | | | |
| DQ2 | 广州东方博物馆 Guangzhou Oriental Museum | | | | |
| DF1 (仿品) (Forgery) | 广州东方博物馆 Guangzhou Oriental Museum | | | | |
| Z1 | 中国科学院高能物理研究所 Institute of High Energy Physics – Chinese Academy of Science | | | | |

（续表）

| 代号 # | 来源单位 Source | 标本图片 Picture | 显微图片 Microscopic Image | | | |
|---|---|---|---|---|---|---|
| Z6 | 中国科学院高能物理研究所 Institute of High Energy Physics – Chinese Academy of Science | | | | | |
| Z10 | 中国科学院高能物理研究所 Institute of High Energy Physics – Chinese Academy of Science | | | | | |
| Z12 | 中国科学院高能物理研究所 Institute of High Energy Physics – Chinese Academy of Science | | | | | |
| J1 | 景德镇陶瓷考古研究所 Jingdezhen Ceramics Archeology Research Institute | | | | | |
| J2 | 景德镇陶瓷考古研究所 Jingdezhen Ceramics Archeology Research Institute | | | | | |

(续表)

| 代号 # | 来源单位 Source | 标本图片 Picture | 显微图片 Microscopic Image | | | | |
|---|---|---|---|---|---|---|---|
| J3 | 景德镇陶瓷考古研究所 Jingdezhen Ceramics Archeology Research Institute | | | | | | |
| J4 | 景德镇陶瓷考古研究所 Jingdezhen Ceramics Archeology Research Institute | | | | | | |
| J6 | 景德镇陶瓷考古研究所 Jingdezhen Ceramics Archeology Research Institute | | | | | | |
| J7 | 景德镇陶瓷考古研究所 Jingdezhen Ceramics Archeology Research Institute | | | | | | |

(续表)

| 代号 # | 来源单位 Source | 标本图片 Picture | 显微图片 Microscopic Image | | |
|---|---|---|---|---|---|
| J8 | 景德镇陶瓷考古研究所 Jingdezhen Ceramics Archeology Research Institute | | | | |
| J9 | 景德镇陶瓷考古研究所 Jingdezhen Ceramics Archeology Research Institute | | | | |
| J11 | 景德镇陶瓷考古研究所 Jingdezhen Ceramics Archeology Research Institute | | | | |
| J12 | 景德镇陶瓷考古研究所 Jingdezhen Ceramics Archeology Research Institute | | | | |
| J14 | 景德镇陶瓷考古研究所 Jingdezhen Ceramics Archeology Research Institute | | | | |

（续表）

| 代号 # | 来源单位 Source | 标本图片 Picture | 显微图片 Microscopic Image | | | | |
|---|---|---|---|---|---|---|---|
| J15 | 景德镇陶瓷考古研究所 Jingdezhen Ceramics Archeology Research Institute | | | | | | |
| T2 | 泰国曼谷大学东南亚陶瓷博物馆 Southeast Asian Ceramics Museum, Bangkok University | | | | | | |
| T3 | 泰国曼谷大学东南亚陶瓷博物馆 Southeast Asian Ceramics Museum, Bangkok University | | | | | | |

（续表）

| 代号 # | 来源单位 Source | 标本图片 Picture | 显微图片 Microscopic Image | | |
|---|---|---|---|---|---|
| T4 | 泰国曼谷大学东南亚陶瓷博物馆 Southeast Asian Ceramics Museum, Bangkok University | | | | |
| T5 | 泰国曼谷大学东南亚陶瓷博物馆 Southeast Asian Ceramics Museum, Bangkok University | | | | |
| T6 | 泰国曼谷大学东南亚陶瓷博物馆 Southeast Asian Ceramics Museum, Bangkok University | | | | |
| T7 | 泰国曼谷大学东南亚陶瓷博物馆 Southeast Asian Ceramics Museum, Bangkok University | | | | |
| T8 | 泰国曼谷大学东南亚陶瓷博物馆 Southeast Asian Ceramics Museum, Bangkok University | | | | |

（续表）

| 代号 # | 来源单位 Source | 标本图片 Picture | 显微图片 Microscopic Image | | |
|---|---|---|---|---|---|
| T10 | 泰国曼谷大学东南亚陶瓷博物馆 Southeast Asian Ceramics Museum, Bangkok University | | | | |
| T12 | 泰国曼谷大学东南亚陶瓷博物馆 Southeast Asian Ceramics Museum, Bangkok University | | | | |
| T13 | 泰国曼谷大学东南亚陶瓷博物馆 Southeast Asian Ceramics Museum, Bangkok University | | | | |
| T14 | 泰国曼谷大学东南亚陶瓷博物馆 Southeast Asian Ceramics Museum, Bangkok University | | | | |
| T16 | 泰国曼谷大学东南亚陶瓷博物馆 Southeast Asian Ceramics Museum, Bangkok University | | | | |

（续表）

| 代号<br># | 来源单位<br>Source | 标本图片<br>Picture | 显微图片<br>Microscopic Image | | | |
|---|---|---|---|---|---|---|
| T17 | 泰国曼谷大学东南亚陶瓷博物馆<br>Southeast Asian Ceramics Museum, Bangkok University | | | | | |
| T18 | 泰国曼谷大学东南亚陶瓷博物馆<br>Southeast Asian Ceramics Museum, Bangkok University | | | | | |
| T19 | 泰国曼谷大学东南亚陶瓷博物馆<br>Southeast Asian Ceramics Museum, Bangkok University | | | | | |
| T20 | 泰国曼谷大学东南亚陶瓷博物馆<br>Southeast Asian Ceramics Museum, Bangkok University | | | | | |
| T21 | 泰国曼谷大学东南亚陶瓷博物馆<br>Southeast Asian Ceramics Museum, Bangkok University | | | | | |

（续表）

| 代号 # | 来源单位 Source | 标本图片 Picture | 显微图片 Microscopic Image | | |
|---|---|---|---|---|---|
| T23 | 泰国曼谷大学东南亚陶瓷博物馆 Southeast Asian Ceramics Museum, Bangkok University | | | | |
| T25 | 泰国曼谷大学东南亚陶瓷博物馆 Southeast Asian Ceramics Museum, Bangkok University | | | | |
| T26 | 泰国曼谷大学东南亚陶瓷博物馆 Southeast Asian Ceramics Museum, Bangkok University | | | | |
| T27 | 泰国曼谷大学东南亚陶瓷博物馆 Southeast Asian Ceramics Museum, Bangkok University | | | | |

（续表）

| 代号<br># | 来源单位<br>Source | 标本图片<br>Picture | 显微图片<br>Microscopic Image | | | | |
|---|---|---|---|---|---|---|---|
| T28 | 泰国曼谷大学东南亚陶瓷博物馆<br>Southeast Asian Ceramics Museum, Bangkok University | | | | | | |
| T30 | 泰国曼谷大学东南亚陶瓷博物馆<br>Southeast Asian Ceramics Museum, Bangkok University | | | | | | |
| T31 | 泰国曼谷大学东南亚陶瓷博物馆<br>Southeast Asian Ceramics Museum, Bangkok University | | | | | | |
| T33 | 泰国曼谷大学东南亚陶瓷博物馆<br>Southeast Asian Ceramics Museum, Bangkok University | | | | | | |
| T34 | 泰国曼谷大学东南亚陶瓷博物馆<br>Southeast Asian Ceramics Museum, Bangkok University | | | | | | |

（续表）

| 代号<br>#  | 来源单位<br>Source | 标本图片<br>Picture | 显微图片<br>Microscopic Image | | | |
|---|---|---|---|---|---|---|
| T35 | 泰国曼谷大学东南亚陶瓷博物馆<br>Southeast Asia Ceramics Museum, Bangkok University | | | | | |
| T36 | 国曼谷大学东南亚陶瓷博物馆<br>Southeast Asian Ceramics Museum, Bangkok University | | | | | |
| N2 | 内蒙古包头市博物馆<br>Inner Mongolia Baotou Museum | | | | | |

附录 Appendix | 249

（续表）

| 代号<br>#  | 来源单位<br>Source | 标本图片<br>Picture | 显微图片<br>Microscopic Image | | | |
|---|---|---|---|---|---|---|
| N3 | 内蒙古包头市博物馆<br>Inner Mongolia Baotou Museum | | | | | |
| N4 | 内蒙古包头市博物馆<br>Inner Mongolia Baotou Museum | | | | | |
| N5 | 内蒙古包头市博物馆<br>Inner Mongolia Baotou Museum | | | | | |
| N6 | 内蒙古包头市博物馆<br>Inner Mongolia Baotou Museum | | | | | |

(续表)

| 代号 # | 来源单位 Source | 标本图片 Picture | 显微图片 Microscopic Image | | | |
|---|---|---|---|---|---|---|
| N9 | 内蒙古包头市博物馆 Inner Mongolia Baotou Museum | | | | | |
| N10 | 内蒙古包头市博物馆 Inner Mongolia Baotou Museum | | | | | |
| N12 | 内蒙古包头市博物馆 Inner Mongolia Baotou Museum | | | | | |
| H1 | 台湾鸿禧美术馆 The Chang Foundation | | | | | |
| H2 | 台湾鸿禧美术馆 The Chang Foundation | | | | | |

(续表)

| 代号<br>#  | 来源单位<br>Source | 标本图片<br>Picture | 显微图片<br>Microscopic Image | | |
|---|---|---|---|---|---|
| H4 | 台湾鸿禧美术馆<br>The Chang Foundation | | | | |
| H5 | 台湾鸿禧美术馆<br>The Chang Foundation | | | | |
| H7 | 台湾鸿禧美术馆<br>The Chang Foundation | | | | |
| H8 | 台湾鸿禧美术馆<br>The Chang Foundation | | | | |
| B1 [113] | 美国波士顿美术博物馆<br>Museum of Fine Arts, Boston | | | | |

| 代号 # | 来源单位 Source | 标本图片 Picture | 显微图片 Microscopic Image | | | | |
|---|---|---|---|---|---|---|---|
| B2 [113] | 美国波士顿美术博物馆 Museum of Fine Arts, Boston | | | | | | |
| B3 [113] | 美国波士顿美术博物馆 Museum of Fine Arts, Boston | | | | | | |

(续表)

(续表)

| 代号<br># | 来源单位<br>Source | 标本图片<br>Picture | 显微图片<br>Microscopic Image | | | |
|---|---|---|---|---|---|---|
| B5 [113] | 美国波士顿美术博物馆<br>Museum of Fine Arts, Boston | | | | | |
| B7 [113] | 美国波士顿美术博物馆<br>Museum of Fine Arts, Boston | | | | | |
| S3 | 美国旧金山亚洲艺术博物馆<br>Asian Art Museum of San Francisco | | | | | |
| K3 | 美国德克萨斯州金贝尔艺术博物馆<br>Kimbell Art Museum, Fort Worth | | | | | |

(续表)

| 代号<br># | 来源单位<br>Source | 标本图片<br>Picture | 显微图片<br>Microscopic Image | |
|---|---|---|---|---|
| K4 | 美国德克萨斯州金贝尔艺术博物馆<br>Kimbell Art Museum, Fort Worth | | | |
| JS1 | 美国洁蕊堂<br>The Stamen Collection | | | |

# 参考文献
## References

[1] 李辉柄. 青花瓷器鉴定[M]. 北京：紫禁城出版社, 2012: 10.

[2] 耿宝昌. 明清瓷器鉴定[M]. 北京：紫禁城出版社, 1993.

[3] 许之衡. 饮流斋说瓷[M]. 济南：山东画报出版社, 2010: 16.

[4] 曹昭. 格古要论[M]. 北京：中华书局, 2012: 217.

[5] 许之衡. 饮流斋说瓷[M]. 济南：山东画报出版社, 2010: 13.

[6] 许哲. 古陶瓷鉴定方法发展史初探[J]. 北方文物, 2011, (2): 44-47. http://www.cnki.net/KCMS/detail/detail.aspx?QueryID=56&CurRec=1&recid=&filename=BJWW201102010&dbname=CJFD2011&dbcode=CJFQ&pr=&urlid=&yx=.

[7] 戴春燕, 杜锋. 浅谈古陶瓷的鉴定方法[J]. 山东陶瓷, 2005, 28(3): 40-43. http://www.cnki.net/KCMS/detail/detail.aspx?QueryID=8&CurRec=2&recid=&filename=BOWL200503014&dbname=CJFD2005&dbcode=CJFQ&pr=&urlid=&yx=.

[8] 郑乃章, 吴军明, 吴隽, 苗立峰. 古陶瓷研究和鉴定中的化学组成仪器分析法[J]. 中国陶瓷, 2007, 43(5): 52-54. http://www.cnki.net/KCMS/detail/detail.aspx?QueryID=60&CurRec=1&recid=&filename=ZGTC200705014&dbname=cjfd2007&dbcode=CJFQ&pr=&urlid=&yx=.

[9] 吴隽. 古陶瓷科技研究与鉴定[M]. 北京：科学出版社, 2009.

[10] 王维达. 古陶瓷热释光测定年代的研究和进展[J]. 中国科学, 2009, 39(11): 1767-1799. http://www.cnki.net/KCMS/detail/detail.aspx?QueryID=16&CurRec=1&recid=&filename=JEXK200911002&dbname=CJFD2009&dbcode=CJFQ&pr=&urlid=&yx=.

[11] 同[8].

[12] 同[9].

[13] 同[9].

[14] 简虎, 吴松坪, 姚高尚, 熊腊森. 能量色散X射线荧光光谱分析及其应用[J]. 电子质量, 2006(1): 13-15. http://www.cnki.net/KCMS/detail/detail.aspx?QueryID=52&CurRec=1&recid=&filename=DZZN200601006&dbname=cjfd2006&dbcode=CJFQ&pr=&urlid=&yx=.

[15] 梅里亚姆-韦伯斯特辞典. http://www.merriam-webster.com/.

[16] 单淼. 真品赝品雾里看花该信眼学还是科学[N]. 新商报, 2012-8-27(4).

[17] 同[16].

[18] 杭州南宋官窑博物馆. 南宋官窑文集[M]. 北京：文物出版社, 2004.

[19] 牛海荣. 数码风潮终结柯达百年传奇[N]. 国际先驱导报, 2012-01-17. http://ihl.cankaoxiaoxi.com/2012/0117/11118.shtml.

[20] 冯先铭. 中国陶瓷[M]. 上海：上海古籍出版社, 2006.

[21] 中国大百科全书总编辑委员会《考古学》编辑委员会. 中国大百科全书·考古学[M]. 北京：中国大百科全书出版社, 1986.

[22] 同[9].

[23] 同[20].

[24] 李家驹. 陶瓷工艺学[M]. 北京：中国工业出版社, 2010.

[25] 同[24].

[26] 同[20].

[27] 同[20].

[28] 张福康. 中国古陶瓷的科学[M]. 上海：上海人民美术出版社, 2000.

[29] 同[28].

[30] 同[28].

[31] 同 [28].

[32] 同 [28].

[33] 同 [28].

[34] 同 [28].

[35] 顾翼东. 化学词典 [M]. 上海：上海辞书出版社, 1989.

[36] 赵彦钊, 殷海荣. 玻璃工艺学 [M]. 北京：化学工业出版社, 2006.

[37] 同 [24].

[38] 同 [35].

[39] 张金升, 张银燕, 王美婷, 许凤秀. 陶瓷材料显微结构与性能 [M]. 北京：化学工业出版社, 2007.

[40] 同 [39].

[41] 袁世金, 冯涛. 中国百科大辞典 [M]. 北京：华夏出版社, 1990.

[42] 同 [36].

[43] 同 [39].

[44] 贺可音. 硅酸盐物理化学 [M]. 武汉：武汉理工大学出版社, 1995.

[45] 中国冶金百科全书总编辑委员会耐火材料卷编辑委员会. 中国冶金百科全书·耐火材料卷 [M]. 北京：冶金工业出版社, 1997：393.

[46] 同 [44].

[47] 同 [44].

[48] 同 [18].

[49] 同 [44].

[50] 同 [41].

[51] 同 [41].

[52] 同 [41].

[53] 同 [1].

[54] 俞伟超. 考古类型学的理论与实践 [M]. 北京：文物出版社, 1989.

[55] 同 [4].

[56] 同 [24].

[57] 同 [24].

[58] 同 [24].

[59] 同 [24].

[60] 冯先铭. 中国陶瓷 [M]. 上海：古籍出版社, 2001.

[61] 同 [60].

[62] 同 [28].

[63] 同 [28].

[64] 同 [28].

[65] 同 [28].

[66] 同 [28].

[67] 胡东波, 张红燕, 刘树林. 景德镇明代御窑遗址出土瓷器分析研究 [M]. 北京：科学出版社, 2011.

[68] 同 [28].

[69] 同 [28].

[70] 罗宏杰, 李家治, 高力明. 中国古瓷中钙系釉类型划分标准及其在瓷釉研究中的应用 [J]. 硅酸盐通报, 1995(2)：50-53. http://www.cnki.net/KCMS/detail/detail.aspx?QueryID=44&CurRec=1&recid=&filename=GSYT199502010&dbname=CJFD1995&dbcode=CJFQ&pr=&urlid=&yx=.

[71] 同 [28].

[72] 同 [28].

[73] 田英良, 孙诗兵. 新编陶瓷工艺学 [M]. 北京：中国轻工业出版社, 2009.

[74] 李丽霞, 贾茹. 硅酸盐物理化学 [M]. 天津：天津大学出版社, 2010.

[75] 同 [36].

[76] 曹志峰, 赵志强. 重火石光学玻璃化学稳定性的研究 [J]. 长春光学精密机械学院学报, 1987, 31(2)：59-65. http://www.cnki.net/KCMS/detail/detail.aspx?QueryID=64&CurRec=1&recid=&filename=CGJM198702010&dbname=CJFD1987&dbcode=CJFQ&pr=&urlid=&yx=.

[77] 同 [76].

[78] 同 [36].

[79] 冯明良. 玻璃表面之组成, 结构与特性 [J]. 玻璃, 1986(2)：39-42. http://www.cnki.net/KCMS/detail/detail.aspx?QueryID=20&CurRec=1&recid=&filename=BLZZ198602011&dbname=CJFD1986&dbcode=CJFQ&pr=&urlid=&yx=.

[80] 王承遇, 陶瑛. 玻璃的表面结构和性质 [J]. 硅酸盐通报, 1982(1)：24-34. http://www.cnki.net/KCMS/detail/detail.aspx?QueryID=24&CurRec=1&recid=&filename=GSYT198201007&dbname=CJFD1982&dbcode=CJFQ&pr=&urlid=&yx=.

[81] 同 [24].

[82] 同 [24].

[83] 陈尧成, 张志刚, 郭演仪. 历代青花瓷器和青花色料的研究 [J]. 硅酸盐学报, 1978, 6(4)：225-241 http://www.cnki.net/KCMS/detail/detail.aspx?QueryID=40&CurRec=1&recid=&filename=GXYB197804000&dbname=CJFD1979&dbcode=CJFQ&pr=&urlid=&yx=.

[84] http://www.flickr.com/

[85] 同[39].
[86] 同[24].
[87] 王承遇, 陶瑛. 硅酸盐玻璃的风化[J]. 硅酸盐学报, 2003, 31(1): 78-85. http://www.cnki.net/KCMS/detail/detail.aspx?QueryID=28&CurRec=1&recid=&filename=GXYB200301016&dbname=CJFD2003&dbcode=CJFQ&pr=&urlid=&yx=.
[88] 同[36].
[89] 同[18].
[90] 同[44].
[91] 宋中庸, 杰戴B·克雷夫特, 马尔科姆G·麦克拉雷. 铅从釉与玻璃中被酸溶液溶出的机理[J]. 河北陶瓷, 1979(02): 51-55. http://www.cnki.net/KCMS/detail/detail.aspx?QueryID=68&CurRec=1&recid=&filename=HBTC197902009&dbname=CJFD1979&dbcode=CJFQ&pr=&urlid=&yx=.
[92] 同[87].
[93] 同[18].
[94] 朱铁权, 王昌燧, 毛振伟, 李立新, 黄烘. 我国古代不同时期铅釉陶表面腐蚀物的分析研究[J]. 光谱学与光谱分析, 2010, 30(01): 266-269. http://www.cnki.net/KCMS/detail/detail.aspx?QueryID=32&CurRec=1&recid=&filename=GUAN201001072&dbname=CJFD2010&dbcode=CJFQ&pr=&urlid=&yx=.
[95] 戴树桂. 环境化学[M]. 北京: 高等教育出版社, 2006.
[96] 同[95].
[97] 同[95].
[98] 马燕如. 我国水下考古发掘陶瓷器的脱盐保护初探[J]. 博物馆研究, 2007, 97(1): 85-89. http://www.cnki.net/KCMS/detail/detail.aspx?QueryID=36&CurRec=1&recid=&filename=BWYJ200701019&dbname=CJFD2007&dbcode=CJFQ&pr=&urlid=&yx=.
[99] 同[42].
[100] 同[83].
[101] 同[9].
[102] 同[80].
[103] bbs.china.com.cn.
[104] 读者. 2011, 21.
[105] www.posterlounge.co.uk.
[106] 读者. 2010, 6.
[107] 同[106].
[108] http://www.flickr.com/photos/mackro/2840868525/sizes/z/in/photostream/.
[109] 读者. 2013, 1.
[110] http://farm1.staticflickr.com/44/134454417_b852cb8099_o.jpg.
[111] http://www.flickr.com/photos/snowpeak/3762193048/lightbox/.
[112] http://www.flickr.com/photos/smithsonian/2941525398/in/photostream/lightbox/.
[113] Photograph © [2013] Museum of Fine Arts, Boston.

Wine jar with design from a popular drama
Chinese, late Yuan dynasty, mid-14th century
Porcelain, Jingdezhen ware, with underglaze cobalt blue
27.8 x 21 cm (10 15/16 x 8 1/4 in.)
Museum of Fine Arts, Boston
Bequest of Charles Bain Hoyt—Charles Bain Hoyt Collection
50.1339

Meiping-shaped vase with blue-and-white decoration of peony scrolls
Chinese, Yuan dynasty, late 14th century
Porcelain, Jingdezhen ware
38 cm (14 15/16 in.)
Museum of Fine Arts, Boston
Gift of Mr. and Mrs. F. Gordon Morrill
1974.480

Meiping-shaped vase with blue-and-white decoration of Daoist immortals
Chinese, Yuan dynasty, mid-14th century
Porcelain, Jingdezhen ware
Overall: 42.4cm (16 11/16in.)
Other (diameter of foot): 13 cm (5 1/8 in.)
Museum of Fine Arts, Boston
Harriet Otis Cruft Fund and Seth Kettell Sweetser Fund
24.113

Trapezoidal blue and white shard with lotus pond design and foot

ring
Chinese, Yuan dynasty, early 13th – late 14th century
Porcelain; Jingdezhen ware
Overall: 12.5 x 8 cm (4 15/16 x 3 1/8 in.)
Museum of Fine Arts, Boston
Gift of Steve and Barbara Gaskin
2006.1823

Trapezoidal blue and white shard with phoenix and peony scrolls
Chinese, Yuan dynasty, 1271 – 1368
Porcelain; Jingdezhen ware
Overall: 8.5 x 11 cm (3 3/8 x 4 5/16 in.)
Museum of Fine Arts, Boston
Gift of Steve and Barbara Gaskin
2006.1825